Econophysics and Physical Economics

Econophysics and Physical Economics

Peter Richmond, Jürgen Mimkes,
and Stefan Hutzler

OXFORD
UNIVERSITY PRESS

OXFORD
UNIVERSITY PRESS

Great Clarendon Street, Oxford, OX2 6DP,
United Kingdom

Oxford University Press is a department of the University of Oxford.
It furthers the University's objective of excellence in research, scholarship,
and education by publishing worldwide. Oxford is a registered trade mark of
Oxford University Press in the UK and in certain other countries

First Edition published in 2013

Impression: 1

Published in the United States of America by Oxford University Press
198 Madison Avenue, New York, NY 10016, United States of America

British Library Cataloguing in Publication Data

Data available

Library of Congress Control Number: 2013937930

ISBN 978–0–19–967470–1

Printed and bound by
CPI Group (UK) Ltd, Croydon, CR0 4YY

Preface

Financial assets are vital parts of the global economy, and an understanding of the origin and nature of price movements is crucial to management of risk. The sums of money involved can be measured in trillions of euros, dollars, or yen—choose any currency you wish, the figures are huge. Time investments correctly and it is possible to make millions of dollars. Robert Merton of Harvard has pointed out that a dollar invested in 1926 in US Treasury bills would have become 14 dollars by 1996. The same dollar invested in the S&P index would have grown to 1,370 dollars. However if perfect timing each month had been possible and the money switched between these two investment routes the total by 1996 would have become 2,296,183,456 dollars! Undoubtedly, and despite the ongoing financial crisis, the total in 2012 would have become at least ten times this value, as may be estimated by considering only the main three peaks and troughs in the index since 1996. Really clever people might have achieved much more by riding the smaller waves within the booms and crashes!

One should not therefore be surprised that people dream about such gains. Birth dates are widely used to fill out lottery tickets and football pools (and it is probably true that the people who spend money in this way are the people who are least able to afford to lose their stake).

During a recent visit to the London School of Economics, Queen Elizabeth II of England is reported to have asked of the economists why they were not able to predict the financial crash of 2008–9. The news item did not give details of any answer. One suspects no answer was provided. Given this situation it is not surprising that a frequently told joke is that economists disagree with each other so often that they make astrologists look good.

For some years now, many economists have agreed over economic theory. US President Richard Nixon, defending deficit spending against the conservative charge that it was 'Keynesian', is reported to have replied, 'We're all Keynesians now.' According to E. Roy Weintraub, economics professor at Duke University and associate editor of *History of Political Economy,* Nixon should have said 'We're all neoclassical economists now, even the Keynesians, because mainstream economics, the subject taught to students today, is neoclassical economics.' This theory is built on assumptions, such as the rationality of economic agents, the invisible hand, and market efficiency, that have become dogma. Physicist J-P Bouchaud has recently recounted the tale of an economist who once told him, to his bewilderment: 'These concepts are so strong that they supersede any empirical observation.'

That being said, some economists are now beginning to question these approaches and recognize that an economy is better described as a complex system whose properties emerge from the interactions of its individual agents. These agents do not always have complete information on which to make decisions; they cannot always be assumed to be the same ubiquitous 'rational agent' of the neoclassical theory. But there remain strong schools committed to the neoclassical theory which is still widely taught in our universities.

It is impossible to imagine such a situation arising in the natural sciences. Physics and the other natural sciences are empirical disciplines. If empirical observation is incompatible with any model, it is assumed that the model or basic principles are wrong and so must be modified or even abandoned. So many accepted ideas have been proven wrong in the history of physics that physicists can be intensely critical of both their own models and those of others. But why are physicists interested in economics and finance?

Physicists, as we shall see in the introductory chapter, have since the renaissance period, been involved in the development and application of the subject, and if one includes the writings of Aristotle, even since the beginning of science itself in ancient Greece. Leaving it around the time of the industrial revolution, a few returned at the turn of the twentieth century to study the subject as the tools of thermodynamics and statistical mechanics were sharpened. The direct involvement of many physicists in finance really got underway in the 1990s as computers took over the recording of financial activity giving rise to vast amounts of data for analysis and, in the earlier chapters, we shall present some of the work that has been done.

In the second part of the book we outline an approach to economics in a manner analogous to thermodynamics and statistical physics, developed by the nineteenth century scientist Josiah Willard Gibbs and others. The resulting laws apply to assemblies of large numbers of interacting molecules for which experiments may be not be repeated sufficiently enough to obtain any but the most probable results. We argue that similar laws apply to assemblies of interacting economic agents for which repeatable experiments are also not always possible. The theory leads naturally to an understanding of a range of financial and economic phenomena. One central issue, namely that of non-equilibrium, is also discussed by drawing on recent ideas developed to explore the phenomenon in physical systems, which leads to new insights into the distribution functions of the interacting agents. It is our view that this approach, which combines both theory and empiricism, offers scope for further development and application.

There have been many books published by the economic and mathematical community on quantitative finance in recent years. However, these can be difficult to read for a person trained in natural science; the mathematics can also sometimes seem to obscure the basic science involved.

Some important books and numerous articles have been published by members of the physics community in the decade or so since the very first conference concerned with the 'Application of Physics to Financial Analysis' (APFA) was held in Dublin in 1999. In particular, books by Rosario Mantegna and Gene Stanley (2000), Jean-Philip Bouchaud and Marc Potters (2000), and Johannes Voit (2001) sought to bring 'econophysics'—a name coined by Gene Stanley—into the mainstream. The recent book by Sinha *et al.* (2011) extends the scope of these books by also addressing progress in understanding the nature of wealth distributions and topics such as game and network theory.

It was at the time of the first APFA conference that one of us (PR) whilst in Trinity College Dublin, gave a few lectures on the topic, which over the next few years developed into a short course given to undergraduates as an option in their final year. This was supplemented by a short project for some of the course members. As lecture notes were prepared, we became aware that none of the existing books on the subject proved ideal in the sense of being a pedagogic text for students new to the topic. Our overall aim here was to provide a basic introduction to the application of physics to economic and financial systems for such young students, based on both the course given over the past decade in Dublin and complementary lectures delivered by JM at the University of Paderborn and SH at Trinity College Dublin.

The Dublin lectures contributed to a new generation of Irish physics students choosing to take up a career in the financial service industry as the 'Celtic Tiger' began to roar and the Irish economy was booming. Sadly, policy mistakes within government and the banking community and affecting the construction industry ensured the scale of the current crisis in Ireland was greater than it otherwise might have been. However, the nation, and indeed the rest of the world, will in time come out of this crisis. Hopefully by then—ready to face up to other future crises— we shall have a properly trained group of expert advisers within both banking and government, not just in Ireland but across the world.

Acknowledgments

The content owes a lot to others who have read and commented on the manuscript and the Trinity College Dublin postdoctoral and graduate students Przemek Repetowicz, Lorenzo Sabatelli, Ricardo Coelho, and Stephen Hardiman who contributed to some of the original work presented. The latter made numerous comments on early drafts and, arguably of greater importance, he has produced the bulk of the figures and data analysis used in the text, for which we are very grateful. Further data analysis was carried out by the graduate students Mike Sexton and David Whyte who also assisted in proofreading the final manuscript.

Two of the authors (PR and JM) are grateful to European COST Action MP0801— Physics of Competition and Conflicts—for financial support during the writing of this text. (For a summary of the activities of this COST action see *Advances in Complex Systems*, **15**, issue supp01 (2012)).

P. Richmond, J. Mimkes, S. Hutzler
Norwich, Paderborn, Dublin, June 2013

Contents

Frequently used symbols

The numbers refer to the *chapter* section where a symbol is first introduced.

k_B	1.4	Boltzmann constant
S	1.4	entropy
$s(t)$	2.1	price as a function of time t
$r(t, \delta t)$	2.1	log price return as a function of time t and time interval δt
$\sigma_r(t, \delta t)$	3.1	instantaneous volatility of log returns
$G(\tau)$	2.1	autocorrelation function
$R(\tau)$	2.1	normalized autocorrelation function
$p(x, t)$	3.1	probability density function
$C_<(X)$	3.1	cumulative distribution function
$C_>(X)$	3.1	complementary cumulative distribution function
$m_1 = m$	3.1	first moment, mean
m_n	3.1	n-th moment
σ	3.1	standard deviation
σ^2	3.1	variance
v_d	3.2	drift velocity
D	3.2	diffusion coefficient
D_n	7.1	generalized diffusion coefficients
P_u	15.1	profit/surplus
Y	15.1	income
C	15.1	costs
S_e	15.3	economic entropy
λ	15.4	dither

1
Introduction

Send your grain across the seas, and in time you will get a return. Divide your merchandise among seven ventures, eight maybe, since you do not know what disasters may occur on earth.

Ecclesiastes 11:1-2

1.1 Physicists, finance, and economics

Is physics what physicists do? Perhaps, but of all the sciences, physics is also arguably the only science that takes upon itself to explore those areas of the universe where the fundamental laws are unknown.

People are frequently surprised to learn that physicists are applying their subject to finance and economics. One young school leaver who was thinking about studying physics said to one of us a couple of years ago that her 'dad would be relieved to know that the subject I want to study has some practical application'. Many colleagues of physicists who are currently looking at problems in economics, finance, and sociology still sometimes comment that these subjects do not fall within the domain of physics. History teaches us otherwise. Whilst Aristotle, Copernicus, Newton, Huygens, Pascal, and Halley are all well known for their studies of natural phenomena, they are also remembered for important contributions that led to greater understanding and advances in economics and finance, as we shall see below.

Figure 1.1 The writings of Aristotle (384–322 BC) were important not only for the development of the natural sciences but also for theories of goods exchange and wealth creation. (©iStockphoto.com/sneska).

Figure 1.2 Nicolaus Copernicus (1473–1543) is famous for his development of the heliocentric theory, placing the sun at the centre of our planetary system. He also made important contributions to a theory of money.
(©iStockphoto.com/labsas).

Aristotle (384–322 BC, Figure 1.1) is known for many things. He laid the foundations of logic. He also laid the foundations of physics by suggesting that nature is composed of things that change and that studying these changes is useful. Indeed his treatise was called $\varphi\nu\sigma\iota\kappa\alpha$ [physica] from which the word physics originated! His thoughts on economics arose from his ideas on politics and society and these ideas prevailed into the twentieth century (Schumpeter, 1954).

Aristotle argued that in most societies, goods and services are exchanged initially via bartering. A problem arises when one person wants what another has, but does not possess anything that this other person wants in return. The introduction of money, initially in the form of metals such as gold and silver, with an intrinsic value, helps to overcome this situation. The exchange now is not between two different products, but between one product and something of equivalent value. The idea of precious metals as a store of value was firmly entrenched in the minds of the Spanish as they plundered South America and shipped gold back to Spain. Even in 2012 the value of gold benefits from a global financial crisis!

Nicolaus Copernicus (1473–1543, Figure 1.2) is universally acknowledged as the founder of modern astronomy. *De Revolutionibus Orbium Coelestium Libri VI* (Six Books Concerning the Revolutions of the Heavenly Spheres), published in 1543, is one of the great books of science, such as Newton's *Principia* and Darwin's *Origin of Species*.

But Copernicus also achieved great reputation for his work on economics as an adviser to both the Prussian State Parliament and the King of Poland. Copernicus was acutely aware of the economic and social distress caused by wartime inflation. In 1522, he prepared the first draft of a report on the economic problems that arise from debasing the currency. As a result he set out a few basic rules for the issue and maintenance of a sound currency. The final report was used by King Sigismund I of Poland during negotiations for a monetary union between Prussia and Poland in 1528.

Figure 1.3 Sir Isaac Newton (1642–1727) is undoubtedly one of the most influential figures in physics. As Master of the Mint he wrote several reports on the value of gold and silver in various European coins.
(©iStockphoto.com/Tony Bagget).

Prior to this time, most contributions to economics and finance were based on moral interpretations by medieval theologians of previous work of Aristotle. They were concerned with what ought to happen rather than with what actually did happen. Copernicus chose to take a purely empirical and pragmatic approach. Rather than appealing to a priori principles or being guided by a philosophical argument, Copernicus developed his ideas in the manner of a physicist, entirely from empirical studies of economic effects.

It is acknowledged today that the work of Copernicus could have been used as a basis for the assessment of the social and economic problems that occurred even during the mid-twentieth century as well as those of the early sixteenth century (Taylor, 1955).

An interesting assessment of the contributions of Sir Isaac Newton (1642-1727, Figure 1.3) to currency issues and finance can be found in the paper by Shirras and Craig (1945). For thirty one years, Newton was first Warden and then Master of the Royal Mint in London. Writing in 1896, W.A. Shaw notes 'Mathematician and philosopher as he was, it is still a matter for genuine admiration to note the unobtrusive skill and lucidity and modesty with which Newton put himself abreast of the keenest coin traffickers of his time' (Shaw, 1896).

Christiaan Huygens (1629–95, Figure 1.4) is widely known for important work in optics. Less well referenced is his work on games of chance. Using John Graunt's (London) book, *Natural and Political Observations Made upon the Bills of Mortality* published in 1662, Huygens constructed a mortality curve. A detailed discussion of issues of life expectancy is found in an exchange of letters with his younger brother Lodewijk. Christiaan Huygens' work may be seen as setting the foundation for the application of the theory of probability to insurance and annuities (Dahlke *et al.*, 1989).

In 1662 Huygens also published on the mechanics of 'Newton's cradle' (Hutzler *et al.* 2004), pictured on the front cover of this book, and discussed it in terms of

Figure 1.4 Christiaan Huygens (1629–95) established wave optics. He also wrote on probability theory and mortality statistics.
(©iStockphoto.com/ZU_09).

conservation of momentum and kinetic energy. However, it was left to Mandelbrot, 300 years later, to show that a collision process of balls or molecules may serve as a metaphor for the exchange of money between interacting people, or agents. We will return to this in detail when discussing modelling wealth distributions in Section 21.3.

Blaise Pascal (1623–62, Figure 1.5) was a French mathematician, physicist, and religious philosopher who made numerous contributions to science, mathematics, and literature. His earliest work was concerned with the construction of mechanical calculators and he also was interested in fluids, clarifying the concepts of pressure, and vacuum. However, arguably his most important contribution to science arose out of his collaboration with Pierre de Fermat that began in 1654. Together they laid the foundations of probability theory (perhaps motivated by correspondence in 1654 with the writer and gambler Chevalier de Méré, concerning the outcome of throws of dice), now widely used not only in quantum theory but also in economics and social science.

In 1654 Pascal was almost killed when horses pulling his carriage bolted. Convinced that it was God who had saved him, he became a committed Christian and gave up gambling. It was after this life-changing event that he proposed his famous wager that is sometimes used to illustrate game theory: 'How can anyone lose who chooses to become a Christian? If, when he dies, there turns out to be no God and his faith was in vain, he has lost nothing—in fact, he has been happier in life than his non-believing friends. If, however, there is a God and a heaven and hell, then he has gained heaven and his skeptical friends will have lost everything in hell!' (taken from Morris, 1982). In the seventeenth century it would have been difficult, if not impossible, to argue against this wager, and even today, many who believe in God would subscribe to Pascal's view. As the former UK Prime Minister, Margaret Thatcher once said, 'with Christianity one is offered a choice'. Pascal codified the choice in terms of odds and probabilities.

Edmond Halley (1656–1742, Figure 1.6) is well known for his work in astronomy and the discovery of the comet that now bears his name. He is less well-known within the physics community for his work in the area of mortality and life expectancy. In

Figure 1.5 Blaise Pascal (1623–62) made important contributions to the understanding of hydrodynamics. Together with Pierre de Fermat he also established probability theory. (©iStockphoto.com/GoergiosArt).

Figure 1.6 Edmond Halley (1656–1742) was a physicist and astronomer, now best known for the identification of 'Halley's' comet. He also studied life expectancies for the pricing of life annuities.
(©iStockphoto.com/picture).

1693 Halley published an article on life annuities based on an analysis of the age-at-death of citizens in Breslau, a town then in Germany but now renamed Wroclaw and located in Poland. At the time of Halley, the city comprised a relatively closed community with little or no immigration and emigration and the associated data was accurate and complete. Halley's article (Halley, 1693) subsequently enabled the British government to sell life annuities in what was then a novel manner. The price of the annuity could be based on the age of the purchaser rather than being simply a constant! Not surprisingly, Halley's work proved to be a key driver for the later development of actuarial science.

Figure 1.7 The mathematician Adolphe Quetelet (1796–1874) (left) and the philosopher Auguste Comte (1798–1857) (right) both considered the term 'social physics' for a scientific approach to study society. Comte is now regarded as the founder of sociology.
(Adolphe Quetelet: Library of Congress/Science Photo Library, Auguste Comte: Science Photo Library).

Prior to this era of mathematical development, the world was full of soothsayers and oracles. With the Renaissance came independent thinking and a break from the passive belief that our future was a whim of the Gods. Risk was something we ourselves had to understand (see for example Bernstein, 1996) and the physicists described in this section made seminal contributions to the topic at a time when it was emerging as a serious subject. Others, such as the reverend and scholar Malthus and the mathematician Verhulst, were building on this scientific activity and developing ideas of growth dynamics and the limits to growth in social and economic systems.

It should therefore be no surprise that the term *social physics* was introduced during the nineteenth century. Adolphe Quetelet (Figure 1.7) was a Belgian mathematician who first used the term in the context of his statistical studies of 'homme moyen', the average man. However, it was the French philosopher, Auguste Comte (Figure 1.7), who defined social physics as the study of the laws of society and the science of civilization (see for example Ball, 2004). He suggested that social physics would complete the scientific description of the world initiated by Galilei, Newton and others working during the eighteenth and nineteenth centuries: 'Now that the human mind has grasped celestial and terrestrial physics, mechanical and chemical, organic physics, both vegetable and animal, there remains one science, to fill up the series of sciences or observation—social physics. This is what men have now most need of; and this it is the principal aim of the present work to establish.' (Comte, 1856).

However, because he disagreed with the approach of Quetelet, Comte discarded the term socio-physics and began to use the new term 'sociologie' or sociology. In the opening page to his Social Physics, Comte notes: 'The theories of social science are still, even in the minds of the best thinkers, completely implicated with the theologico-metaphysical philosophy; and are..., by [this] fatal separation from all other science, condemned to remain so involved forever.'

Figure 1.8 Louis Bachelier (1870–1946) presented in his PhD thesis of 1900 a random walk model for financial data. Only in the late 1950s was his work rediscovered for financial modelling.

Despite the nineteenth century developments of the kinetic gas theory and statistical mechanics which would have provided tools for physicists to develop 'social physics', most of the physics community turned their attention to other topics that demanded serious inquiry, such as the theory of relativity and gravity, the new atomic physics and quantum mechanics.

The exception was the mathematician and theoretical physicist Louis Bachelier 1870–1946 (Figure 1.8) who in 1900 developed a theory of stock prices, based on ideas of (atomic) Brownian motion and random walks. This will be the topic of Chapter 6. The theory was only taken up again in the latter half of the twentieth century by Fisher Black, a theoretical physicist, together with economist Myron Scholes in the context of a theory of option prices. Unfortunately, Black died before the team was awarded the Nobel Prize for economics.

Over the last two decades the situation has changed markedly, with an ever growing number of physicists turning their attention to what is now called *socio-physics*. The highly recommendable book 'Critical Mass' by Philip Ball (2004) gives an excellent account of both the historical origins of socio-physics (Ball starts with the philosopher Thomas Hobbs (1588–1679), whom he attributes as being 'the first to seek a physics of society') as well as comprehensive non-technical summaries of the current state of the field.

Socio-physics describes the application of ideas from physics, in particular from statistical mechanics, but also quantum mechanics and nuclear physics, to problems in sociology (e.g. opinion formation and voting patterns), town-planning (e.g. traffic flow and pedestrian motion), economics, and finance. Physicists working in these areas often refer to 'complex systems' that they are trying to describe. We will now have a closer look at these.

1.2 Complex systems

A financial market is an example of a complex system (Nicolis and Nicolis, 2007, Érdi, 2008); a system that emerges as a result of the dynamic activity of many, many people

Figure 1.9 A tornado is an example of a phenomenon that is highly organised at a large scale (as visible by an outside observer), while random at a local scale (as experienced by somebody in the midst of it).
(©iStockphoto.com/deepspacedave).

who simultaneously engage in financial transactions, causing asset prices to move up and down in what seems to be a random manner. Dealing with systems that have many degrees of freedom, such as those encompassed by the activity of the people in a market, is a non-trivial task.

One of the simplest examples of a complex system, found in the physical world, is that of a fluid. Fluids consist of many individual molecules that interact via quantum mechanical and electromagnetic forces. They frequently produce complex behaviour, which can be either highly organized in the manner of a tornado, or seemingly chaotic as in highly turbulent flow. What is actually seen often depends on the size of the observer. Farmers in the USA frequently see the highly structured phenomena shown in Figure 1.9. However, a fly caught in the wake of such a tornado would no doubt be surprised to learn that it is participating in such a well-defined flow.

Soap bubbles making up a foam (Weaire and Hutzler 1999, 2009), a granular material such as sand, or even a pile of rocks are other examples of a complex system. The detailed (micro-) structure is given in terms of the shape, position and orientation of each particle or bubble and this complete 'set of coordinates' ultimately determines whether the entire system is stable. However, this foam or pile of sand or rocks is more than a collection of granules. The interactions between the different granules give rise to emerging properties that can be seen as the heap of granules collapses, say under gravitational forces, or the foam begins to flow under an applied shear. The avalanches and rearrangements that arise as these changes occur we now know follow well-defined statistical laws.

The earth's climate composed of the atmosphere, the biosphere, oceans, all subject to intra- and extra-terrestrial processes, such as solar radiation, gravitation and the generation of greenhouse gases, form yet another complex system. Even with modern computers we are unable to predict the weather other than in a probabilistic sense. Yet this chaotic system also contains regularities. As winter follows summer and night follows day in a predictable pattern, this is also reflected in systematic seasonal and daily variations of the average temperature.

Biology is a source of many complex systems. The human body consists of numerous interacting organs with numerous functions, dependent on external sources of food, air and water to operate in a well-regulated and living state. At the cellular level, the replication of DNA and formation of proteins are controlled by many simultaneous biochemical processes within the cellular system.

The brain is also a complex system consisting of typically about 10^{12} nerve cells. The collective interaction of these elements allows us to see, hear, speak and perform a whole range of mental tasks. These characteristics can be thought of as emergent features. As artificial intelligence and learning processes are further developed, and perhaps replace the standard logic of algebra, even computers and information networks may be thought to be complex systems.

Finally, one of the most complex systems within our modern networked world is arguably that of society itself—the global society of human beings that form the economy with its managers, employees, consumers, capital goods such as machines, factories, transport systems, natural resources and financial systems.

We begin to see, therefore, that complex systems are characterised by the possibility of organisation or structure formation on various scales. However, defining the correct scales to begin any analysis is not easy. For example in biology we could define a whole hierarchy of levels that range from the atomic level through to the level of the cell and up to the human state. We might think we should begin with quantum mechanics and end up with some description of human behaviour. This would require more information than we probably have at present but even if we had access to all this information it would be beyond the ability of anyone to handle it (see for example the article by Andersen, 1972).

A key point to recall is that any macroscopic description of a system uses only a few variables to characterize and describe the system on some gross length or time scales. Microscopic motion is usually of no concern. For example, a fluid may be described by the local density and velocity fields. However, hidden behind these are the motions of the many molecules that make up the system. Similarly in looking at financial markets, one can choose the variations in asset prices to characterise the macroscopic system. However, behind these fluctuating prices is the global economy, driven by the acts of numerous individuals or businesses engaged in trading, together with external events such as wars or natural disasters, such as earthquakes, eruptions of volcanoes etc.

The essential problem when treating complex systems consists of finding suitable relations that allow prediction of the future evolution of the system. Unfortunately all the microscopic degrees of freedom are normally coupled together and any attempt to accurately predict the macroscopic properties from the microscopic properties is usually a hopeless task.

In summary, although there is no agreed strict definition of complex systems, probably the best way to describe it is to say that complexity is the behaviour of collections of very many interacting units that evolve in time. The interactions, which are often non-linear, are responsible for coherent collective phenomena and so-called emerging properties. These may be described only on a scale much higher than that of individual units, from which they may not be directly deduced. In this sense the whole is more than the sum of its components.

As is clear from the above, the meaning of 'complex' in our context is quite different from its common language use when describing something as 'complicated'. A further distinction should be made to what is called *algorithmic complexity* or Kolmogorov complexity. This refers to the minimum number of bits required to compute a given sequence of characters or numbers (using some universal programming language). Although it is possible to make a link to complex systems as described above, via the mathematical concept of chaotic maps, algorithmic complexity lacks the dynamic and emergent behaviour that characterises complexity in nature (Nicolis and Nicolis, 2007).

1.3   Determinism and unpredictability

The time evolution of physical systems such as a liquid, solid or even the humble gas may in principle be computed using Newtonian mechanics. This theory involves deterministic equations and, as a result, the entire trajectories of all the particles or elements in the system, can in principle be computed if their initial values are known.

The dynamics of a classical physical system with N particles may be formulated in terms of Newton's laws of motion,

$$\mathbf{p}_j = m_j \frac{d\mathbf{x}_j}{dt}; \quad \frac{d\mathbf{p}_j}{dt} = \mathbf{f}_j(\mathbf{x}_1, ..., \mathbf{x}_N, \frac{d\mathbf{x}_1}{dt}, ..., \frac{d\mathbf{x}_N}{dt}) \text{ for } j = 1 \text{ to } N, \qquad (1.1)$$

where $\mathbf{x}_j$ is the position vector of particle j with mass m_j, $d\mathbf{x}_j/dt$ is its velocity, $\mathbf{p}_j$ is its momentum, and $\mathbf{f}_j$ is the force exerted on it by all the other particles. This set of $2N$ equations can then in principle be solved and the positions and momenta for the N particles computed as a function of time t, given the initial conditions, $\mathbf{p}_j(t = t_0)$ and $\mathbf{x}_j(t = t_0)$.

The *components* of position and momentum form a $6N$-dimensional vector $\mathbf{\Gamma} = (\dots q_i \dots; \dots p_i \dots)$ which defines a particular microscopic state of the system. So the entire N-particle system can then be represented by a point in a $6N$ dimensional phase space, spanned by axes corresponding to the $6N$ *degrees of freedom*.

The essential problem of statistical mechanics is to find the relation between the microscopic world described by $\mathbf{\Gamma}$ and the macroscopic world. For simple force functions, such as those linear in the positions (and independent on the velocities), a solution of eqn (1.1) is possible. If the force is non-linear in the coordinates or time dependent—and such forces may well be prevalent in social systems—then one must often resort to computer simulations to obtain a solution.

This deterministic approach to obtaining a solution has been challenged in two distinct ways.

- Quantum mechanics tells us that we cannot know simultaneously both the precise position and momentum of a particle. However, we can still argue that a deterministic character is conserved as a property of the wave function that describes the system.
- We also now know that without absolutely precise knowledge of the initial conditions—something that in general is not possible—predictability of the posi-

tions and momenta over a long time is simply not possible and the behaviour can become chaotic.

Chaos is not observed in simple systems where the governing dynamical equations are linear. For such systems, the superposition principle holds and the sum of two solutions of the governing equations is also a solution. The breakdown of linearity and also therefore of the superposition principle is required for the behaviour of a non-linear system to become chaotic. However non-linearity in itself is not sufficient: a simple pendulum is governed by a non-linear dynamical equation and the solution based on elliptic functions has no randomness or irregularities. Solitons are other examples of regular collective motion found in a system with non-linear couplings. These are stabilised due to the balance between non-linearity and dispersion effects. But as the number of degrees of freedom increases, the potential for deterministic chaos increases. The unpredictability in such a system then arises from the sensitivity to the initial conditions that can, in practice, be measured only approximately.

One important property of Newton's equations (1.1) is that they are invariant with respect to time reversal. Turn the momenta or velocities around in a mechanical system and the system retraces its path and there is no way of deciding the true or preferred route. Within classical mechanics there is no arrow of time. Yet one of the key outcomes of the science of thermodynamics is that, on a macroscopic level, such an arrow of time does exist.

1.4 Thermodynamics and statistical mechanics

One of the greatest scientific achievements that emerged out of the engineering activity of the industrial revolution was thermodynamics. This theory provides us with an understanding of the relationships between energy, work, and heat and is among the most well known theories of the macroscopic world.

A thermodynamic system, i.e. any macroscopic system, may be characterized by only a few measurable parameters such as, in the case of a fluid, pressure, volume, and temperature. A thermodynamic (macro) state is specified by a particular set of values of all these parameters necessary to describe the system. If the thermodynamic state does not change with time, the system is said to be in thermodynamic equilibrium. Once a system reaches that state, it will stay there and all the macroscopic properties will not change any more. This implies that there is, within the laws of thermodynamics, an 'arrow' of time.

Statistical mechanics developed from the question: how can we explain or interpret the laws of thermodynamics in terms of classical mechanics? It seems almost impossible because of the two completely different aspects of the role of time in thermodynamics and classical mechanics. This problem is one that Ludwig Boltzmann (Figure 1.10) grappled with for most of his life before he finally committed suicide in 1906 (see for example Cercignani, 1998).

If we consider an energy-conserving (conservative) N-particle system within a $6N$ dimensional phase space, the point defined by the vector Γ will wander around on a $6N - 1$ dimensional hyper-surface for which the total energy E is constant. The path will never cross itself, because if it did so, that would mean that the system at the crossing point would not know which way to go. Since Newton's equations, eqn (1.1),

Figure 1.10 Ludwig Boltzmann's grave in Vienna. The inscription above his bust reads $S = k \log W$, his famous equation linking entropy S to the number W of possible microstates corresponding to a particular (thermodynamic) macroscopic state. The constant k (written as k_B in this book) is called Boltzmann's constant. (Photograph taken by one of the authors (PR)).

are unique, they always tell how it goes on. According to a theorem of Poincaré, if you begin at a certain point and draw an arbitrary small region around that point, then after a sufficiently long time the path will eventually lead back to that region of phase-space (see for example Uhlenbeck and Ford, 1963).

The *ergodic hypothesis* states that over long periods of time, the time spent by the system in any particular region of the phase space is proportional to the volume of that region. This implies that all accessible regions of phase space are equally probable over a long period of time. So from a mechanical point of view, the motion of the system in phase space is quasi-periodic. There is no sense of the system 'going to equilibrium'. The system never knows whether it is in equilibrium or not! This appears different to the concept of a thermodynamic equilibrium from which a system cannot escape, once it is reached; one might indeed be inclined to say that the two points of view are logically impossible.

The paradox was resolved by Boltzmann, and later by Gibbs, by pointing out that mechanics is microscopic and thermodynamics is macroscopic. Thermodynamic equilibrium is a macroscopic notion and is defined by introducing a small number, M, of macroscopic variables like density, temperature, free energy, and entropy that have a tendency to become uniform across the system under consideration. M is clearly very much smaller than N, the number of entities, such as atoms, that make up the system.

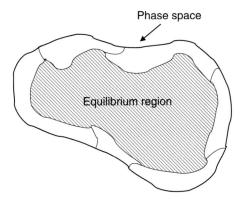

Figure 1.11 Schematic illustration of phase space showing how an *equilibrium* region may coexist with other *non-equilibrium* regions.

To set up our statistical mechanics we introduce the probability density function $p(\mathbf{\Gamma}, t) = p(\ldots q_i \ldots, \ldots p_i \ldots; t)$, such that the probability that the system lies in a small volume element of phase space is given by the probability 'measure' $dP = p(\mathbf{\Gamma}; t)d\Gamma \equiv p(\ldots q_i \ldots, \ldots p_i \ldots)dq_1 \ldots dq_{3N}dp_1 \ldots dp_{3N}$. Here, q_i and p_i are the components of position and momentum of the N particles, respectively. This is an a priori probability that encapsulates our belief about the system, as we shall discuss further at the end of this section. The distribution must be normalized i.e., $\int p(\mathbf{\Gamma}, t)d\Gamma = 1$, where the integral is taken over the entire phase space.

Suppose we now prescribe the values associated with the macroscopic variables. These will correspond to different regions of the underlying $\mathbf{\Gamma}$ space. In this way we see that a set of macroscopic variables correspond to a particular region or maybe regions in the phase space associated with the microscopic variables. Equilibrium statistical mechanics assumes then that there is one region that is overwhelmingly large. This region is then identified with an equilibrium region, as shown in the sketch in Figure 1.11. If the trajectory is located within one of the other smaller regions, then the system is not in equilibrium.

This concept may be illustrated with the following example. Imagine a balloon of volume $V = 1m^3$ containing gas molecules. Normally, all the molecules in the balloon are distributed more or less evenly throughout the entire balloon. From the above description, we can imagine that there is a probability that all the molecules might find themselves in one small element of the space within the balloon, say $\Delta V = 1cm^3$. According to Poincaré's hypothesis, this is certainly possible. If such a state were highly probable, then it is clear that with all the molecules located in one half of the balloon, it would collapse. However, the probability of this happening is extremely tiny. Typically, the probability of a molecule to be confined in a particular region of volume ΔV is about $\Delta V/V \sim 10^{-6}m^3/1m^3 = 10^{-6}$. So the probability of finding all 10^{23} molecules simultaneously in this region is $\sim (10^{-6})^{23}$, which is extremely small.

Given this picture, we can see that when the system is in the equilibrium state, it will almost always stay there. However, in line with Poincaré's theorem, fluctuations occasionally happen that reflect evolution of the system throughout other regions of phase space.

The entropy S of a thermodynamic *macrostate* is defined as

$$S = -k_B \int p(\mathbf{\Gamma}, t) \ln p(\mathbf{\Gamma}, t) d\mathbf{\Gamma}, \tag{1.2}$$

with the normalization condition $\int \rho(\mathbf{\Gamma}, t) d\mathbf{\Gamma} = 1$, and the Boltzmann constant $k_B = 1.381 \times 10^{23} \, JK^{-1}$. In the case of countable, *discrete microstates* r, instead of volume elements $d\mathbf{\Gamma}$ in phase space, the entropy S may be expressed as

$$S = -k_B \sum p_r \ln p_r, \tag{1.3}$$

where p_r is the probability of the microstate p_r. The normalization condition is then $\sum p_r = 1$, where the sum is over all accessible microstates. For an isolated system in equilibrium this leads to

$$S = k_B \ln \Omega, \tag{1.4}$$

where Ω is the number of microstates consistent with the macrostate, as defined by the values of energy, number of particles, and volume of the system (see also the discussion in Section 19.1). Eqn (1.4) is displayed on Boltzmann's gravestone (see Figure 1.10) and arises from assigning equal a priori probabilities $p_r = 1/\Omega$ for all microstates compatible with the above constraints. This reflects that there is no a priori reason that one microstate should be preferred over another for the isolated system in thermal equilibrium.

We now see that the key for extension of this statistical physics approach to other systems in areas, such as economy, sociology, and biology, is to be able to specify the probability for the realization of a certain microscopic state. This may be done either on the basis of the preparation of the complex system or due to a suitable empirical estimation. The intuitive notion of such a probability is clear in games with dice or the tossing of coins, under the assumption that they are fair and unbiased. The probability can be defined empirically by the relative frequency of a particular realization when the game is repeated many times. This so-called 'frequentist' approach to probability theory is not very useful for the characterization of complex systems such as occur in finance, biology, and sociology. Here, we are not able to determine frequencies by repeated experimentation in the scientific sense, nor do we have much information about possible outcomes. The way forward is to interpret the probability as a *degree of belief* that an event will occur. Such an a priori belief is necessarily subjective and may depend on the experience of the observer.

This idea was originally introduced by Thomas Bayes in the eighteenth century in an approach which combines a priori judgments and scientific information in a natural manner and applies to any process, whether or not it is repeatable under constant conditions. We will discuss this in detail in Section 3.3.

The generalization of the method of statistical physics to social, economic, and other systems requires treating these in the sense of Bayes. That is, a system at a microscopic level may be characterized by a probability density distribution, $p(\mathbf{\Gamma}, t = t_0)$ and the evolution of this distribution with time may be represented by a map of

$p(\boldsymbol{\Gamma}, t_0)$ onto the distribution, $p(\boldsymbol{\Gamma}, t)$ at a later time $t > t_0$. In other words, the total probability and hence our degree of belief as embodied in our choice of initial a priori probabilities captured in the initial distribution $p(\boldsymbol{\Gamma}, t_0)$ remains conserved.

It is usually the case that for many particle systems, a complete probability distribution function, $p(\boldsymbol{\Gamma}; t)$ that captures details of all the degrees of freedom is not required. A reduced function is sufficient to obtain all the properties of interest. For example, suppose the microscopic state vector $\boldsymbol{\Gamma}$ encompasses some relevant variables, $\mathbf{x}$ and other redundant or irrelevant variables, $\mathbf{y}$. The probability density, $p(\boldsymbol{\Gamma}; t)$, may then be written as $p(\mathbf{x}, \mathbf{y}; t)$ and the required distribution involving only the relevant variable $\mathbf{x}$ is obtained by integrating over the irrelevant variables, $\mathbf{y}$. Thus $p_r(\mathbf{x}; t) = \int d\mathbf{y} p(\mathbf{x}, \mathbf{y}; t)$. If the initial probability distribution function is normalized, the reduced density function is also normalized, i.e. $\int p_r(\mathbf{x}; t) d\mathbf{x} = 1$.

Averages of arbitrary functions of $\mathbf{x}$ are obtained by summing values of the relevant function $f(\mathbf{x})$, weighted using the probability density function. Thus

$$\langle f(t) \rangle = \iint d\mathbf{x} d\mathbf{y} f(\mathbf{x}) p(\mathbf{x}, \mathbf{y}; t) = \int d\mathbf{x} f(\mathbf{x}) p_r(\mathbf{x}; t). \tag{1.5}$$

In this expression, irrelevant degrees of freedom are hidden in the reduced function $p_r(\mathbf{x}; t)$.

For simplicity, we shall use the notation p rather than p_r for this reduced probability distribution function in the remainder of the book. A detailed discussion of the properties of probability distributions will be subject of Chapter 3.

1.5   Economics, econophysics, and social systems

Economics is traditionally concerned with decision-making and choice. It focuses especially on aspects of choice where resources are not always in abundant supply and where individuals, communities or businesses are required to utilize limited resources in some kind of optimal way. Economics is also concerned with the behaviour of people as they decide what to do under such restrictions. In considering all these issues, the outcomes are generally considered to arise as a result of either intrinsic (endogenous) mechanisms, such as the actions of the system actors or *agents* as they invest, buy or sell assets, or external (exogenous) mechanisms that include shocks such as currency revaluations, the outbreak of war or natural disasters, and so on. The subject is divided into two parts: microeconomics and macroeconomics. The first topic considers the nature and dynamical behaviour of individual elements such as consumers, business owners, management boards, who interact to form, for example, prices; the second considers the aggregate behaviour of such elements in order to describe quantities such as gross domestic product (GDP) and the level of employment.

Economics as a discipline began to take shape during the eighteenth century, a time when the dominant scientific paradigm was classical mechanics, as developed by Newton and Descartes. The French scientist François Quesnay, a physician to Louis XV, proposed the idea of an economy as a social organism driven by mechanical forces–a model of wheels, springs, and cogs. A clock is a sequential mechanical system that is pre-programmed. There is no scope for self-regulation. So in the economic models of Quesnay, advances or the driving forces in agriculture were thought of as the weights and springs. Production was the result of the movements of all these connected mechanisms; prosperity was the outcome of regular or clockwork circulation.

This causal determinism, with no self-regulation or individual freedom, was equated to political systems of absolutism, with individuals simply being functioning elements in a political economic machine. Adam Smith, however, noted that Newton considered that material bodies, as exemplified by the stars in the heavens, form a system moving in state of dynamic equilibrium, determined by invisible forces of gravity. The corresponding physical concept of individuals freely moving in dynamical equilibrium was more in tune with liberal ideas of a free economy and society with a division of political powers. Nevertheless, the description remained essentially a mechanical, deterministic one. The future was somehow already determined by actions in the past. So, for example, all prices and contracts were somehow settled not just now, but at future times, by an invisible hand.

Economists such as Ricardo, Jevons, Walras, Pareto, and Keynes developed these ideas further during the nineteenth and twentieth centuries. The general equilibrium theory, also commonly known as neo-classical economics that rests on the correspondence with classical mechanics became, as the end of the twentieth century approached, the dominant intellectual paradigm within economics. However, whilst neoclassical economics has been valuable in helping understand some aspects of the social and economic world, it has never eliminated other theoretical approaches.

There were aspects of economic systems that could not be accounted for. For example, Pareto noted at the end of the nineteenth century that the distribution of wealth in society appeared to be governed by a distribution that exhibited features that were invariant over time and space. Economic theory could not predict this. (We will report on previous progress on understanding wealth distributions using approaches borrowed from physics in Chapter 21.) It is also very clear that this general equilibrium theory has been unsuccessful in accounting for both the recent and past turbulent behaviour in economic and financial spheres. Neither Keynesian nor monetarist theories have been able to predict real economic behaviour in recent years. Neither can the theories point to any significant success arising from their use by governments in framing policy.

At this point, a physicist would conclude that either the theories are at best incomplete or at worst wrong and at the very least innovation is needed. At the turn of the twentieth century, the ideas of Newton were used to describe the world using classical mechanics. Careful astronomical observations demonstrated that the Newtonian view was only an approximation to the new theory of Einstein. Similar extensive measurements of the physical world at the atomic scale revealed the limitations of Newton's ideas and led to the development of quantum mechanics. Of arguably more interest in the context of this book is the innovation that resolved the difficulties in understanding heat and its link to work and energy, as introduced in the previous section. In this instance the physics and engineering community were forced to recognize that a macroscopic physical system could not be described solely in terms of mechanical variables; a new state function, the so-called entropy, was needed.

The neoclassical theory of economics treats an economic system as a kind of mechanical mechanism (Backhouse, 2002, Cleaver, 2011). However, could it be that an economic system is more akin to a thermodynamic system and that some kind of 'economic entropy' is required? In Chapters 14 to 19 we will develop what we call *physical economics*, in analogy to thermodynamics. The economic entropy that we introduce should be seen as a well-founded replacement for what is called *production* or *utility function* in neoclassical economics.

Entropy introduces fluctuations into the system through transitions between the various microstates and we shall study the nature of these fluctuations in the first part of the book where we examine asset prices.

If the fluctuations are assumed to be Gaussian and the system is effectively in a state of thermodynamic equilibrium, it turns out that the price fluctuations over a time vary with the square root of time, whereas over the same time, the change in the mean price is linearly proportional to time. So one might expect the effect of the fluctuations to lose their importance against the mean price trend for sufficiently large times, when the neoclassical theory might be valid. As we shall see in Section 6.1, sufficiently long time asset prices do fit this model.

For shorter times, the picture is not valid; the fluctuations are not Gaussian and the system is not in equilibrium. Nevertheless, we shall see that we can make progress through methodologies taken over from physics, including the development of expressions for entropy for systems out of equilibrium, see Chapter 20.

Recognition of all the above in itself is not an answer to the question of how physics can contribute to economics. Physics rests for its success on both its methodologies and empirical approach. And in the context of economics, we might imagine the physicist to be in the engine room helping drive a ship, whereas the economist is the captain on the bridge. The definition of the relevant quantities corresponding, for example, to the degrees of freedom, is a task for economic investigation; the physical contribution consists of developing equations that yield relations between the macroscopic economic variables based on general laws. Econophysics seeks not to displace economics, rather it aims to help economists find deeper understandings of complex systems with a large number of degrees of freedom.

And with the advent of modern computers and their implementation in financial markets, gigabytes of information relating to their dynamical behaviour is now becoming available, spanning time scales from the level of ten milliseconds (Preis and Stanley, 2010) through to hundreds of years. Eleven orders of magnitude is more than is available for many other systems, apart from perhaps cosmology. It is not surprising that Gene Stanley coined the name *econophysics* and placed the study of financial markets in the mainstream of physics.

In the same spirit now, that substantial data is becoming available for social phenomena from websites, such as Twitter, Facebook and other internet communications (Caldarelli, 2007, Hardiman *et al.*, 2009), we shall likely see new insights emerge into social networks giving a new renaissance to the topic of sociophysics initiated so long ago by Comte (see also Chapters 13 and 22). However, we begin now our journey with an introduction to properties of financial data (Chapter 2) and continue with the development of some mathematical tools (Chapters 3 to 5), before examining the nature of the dynamics of financial asset prices (Chapters 6 to 12).

2
Reading financial data

Les sottises des ignorans prennent autant de temps a lire, que les bonnes choses des doctes.
The nonsense of fools takes as much time to read as the good things of the learned.
From a letter by Constantijn Huygens to René Descartes.

There is now an enormous wealth of financial data available. Companies publish financial information in the form of accounts. The reporting period varies from country to country. In the US, such reports are required quarterly, while in the UK, companies are required to report twice each year and give details of their financial performance. The data within company reports is concerned with the fundamental health of a company and spans income from sales, expenditure on salaries, capital investment, overall profit or earning, dividend payouts to shareholders, and transfers to the company reserves. These reports together with periodic meetings with company directors are the greatest concern of *fundamental investors*, as opposed to *technical traders* who rely more on the asset price movements over time, rather than fundamental data. In this chapter we shall first describe the characteristics and general way data may be analysed. We then briefly describe these two different styles or approaches to investing.

2.1 Financial price data

Since the nineteenth century, financial transactions have been recorded on a daily basis, and since 1984, the sampling rate of transaction values has been made on a minute-by-minute basis. With the advent of modern computing technology, every transaction is now being recorded 'tick-by-tick', hence the term *tick data*. As a result, giga-amounts of data are now available. In his renowned 1963 study of cotton prices (Mandelbrot, 1963) that revealed for the first time a power law in the probability distribution function for price returns, the mathematician Benoît Mandelbrot used 2000 data points. In 1999, Gene Stanley and co-workers used 40 million points in a study of price fluctuations of 1000 companies quoted on the three major US stock markets: NSYE, AMEX, and NASDAQ (Gopikrishnan *et al.*, 1998).

Modern computers have not only allowed for more effective recording of prices but have also opened the way for more detailed analysis using tools of modern mathematics and physics. This is leading to a much greater understanding of market dynamics – as well as assisting applied studies of option pricing and portfolio management – as we discuss later in Chapters 8 and 11 respectively.

2.1.1   Price scales, time scales, and trading time

In physics, reference units are maintained and improved by selected laboratories. This has led to the science of metrology. In finance, the quoted prices of stocks are generally in the form of currencies that fluctuate relative to other currencies. Moreover, currencies can change not only relative to other currencies, but also relative to assets such as gold. They are also affected by inflation, economic growth, or contraction (recession).

The transactions (events) also occur at random times with random intensity. Physical time is well-defined. However, in finance, time scales take on an extra dimension. Although global markets are now active over the entire 24 hours of each day, there are still periods such as weekends and holidays where all the markets are closed and we really have no idea how to model such closed markets. For example, what is the effect of price sensitive news that arrives over such a closed period? An interesting variation on this is the continuing activity of an online betting market during half-time intervals of soccer matches (see Section 13.1). This is despite there being no news, as the players are off the pitch!

For currency transactions, trading time is (again bar weekends and holidays) equal to physical time. However, closure–closure studies show that the variance between successive days during the week can be up to 20% lower than similar values across weekends (Mantegna and Stanley, 2000).

Market activity is often implicitly assumed to be uniform during trading hours but data shows that both the trading volume and number of contracts made vary. With tick data, it is possible to define time in terms of transactions. However, the volume of transactions remains as a random variable.

In general, care should thus always be taken when comparing different results published in the literature.

2.1.2   Characterizing price fluctuations

Let us now have a closer look at various ways of dealing with time fluctuations of financial data.

Suppose $s(t)$ is the price of an asset at time t. The simplest way to evaluate price fluctuations is to consider the change of the price over a time δt:

$$\delta s(t, \delta t) = s(t + \delta t) - s(t) \tag{2.1}$$

This has the merit of being simple and the price changes are additive,

$$\delta s(t, \delta t_1 + \delta t_2) = \delta s(t + \delta t_1, \delta t_2) + \delta s(t, \delta t_1). \tag{2.2}$$

One difficulty with this approach is, as we implied above, that the fluctuations are affected by changes in prices due to economic changes that occur during periods of inflation or recession. An alternative approach might be to incorporate some kind of inflation factor, $f(t)$, and use the renormalized price changes $\delta \tilde{s}(t, \delta t)$, defined as

$$\delta \tilde{s}(t, \delta t) = f(t + \delta t)s(t + \delta t) - f(t)s(t). \tag{2.3}$$

Of course the problem with this approach is that there is no obvious choice for the function $f(t)$. Another approach that is intuitively more reasonable is to compute the *return*, $r_0(t, \delta t)$,

$$r_0(t, \delta t) = \frac{\delta s(t, \delta t)}{s(t)} = \frac{s(t + \delta t)}{s(t)} - 1. \tag{2.4}$$

This gives the *relative* increase (profit) or decrease (loss) during the period, δt. The disadvantage with this approach is that returns defined in this way are not additive. They are multiplicatively linked, viz

$$[r_0(t, \delta t_1 + \delta t_2) + 1] = [r_0(t + \delta t_1, \delta t_2) + 1][r_0(t, \delta t_1) + 1]. \tag{2.5}$$

This problem can be resolved by using the logarithm of the price and defining the *log-price return*, $r(t, \delta t)$, as

$$r(t, \delta t) = \ln s(t + \delta t) - \ln s(t) = \ln \frac{s(t + \delta t)}{s(t)}. \tag{2.6}$$

The log-price returns are now seen to be additive,

$$r(t, \delta t_1 + \delta t_2) = r(t + \delta t_1, \delta t_2) + r(t, \delta t_1). \tag{2.7}$$

The incorporation of an inflation factor, $f(t)$, leads to

$$\ln \frac{f(t + \delta t)s(t + \delta t)}{f(t)s(t)} = r(t, \delta t) + \ln \frac{f(t + \delta t)}{f(t)}. \tag{2.8}$$

We see that for $f(t) = f(t+\delta t)$ the second term on the RHS is zero; the use of log-price returns takes account of inflation that remains constant over time.

Furthermore, if δt is small and the associated price change is also small, $|\delta s(t, \delta t)| \ll s(t)$, then we see from Eqns (2.4) and (2.6) that $r(t, \delta t) \sim r_0(t, \delta t)$.

A common measure of fluctuations is the absolute value of the log-price returns, $|r(t, \delta t)|$. This is closely related to the *instantaneous volatility*, defined as

$$\sigma_r(t, \delta t) = \sqrt{r(t, \delta t)^2 - \langle r(t, \delta t) \rangle^2}, \tag{2.9}$$

where $\langle r(t, \delta t) \rangle$ is the time average of the log price return,

$$\langle r(t, \delta t) \rangle = \lim \frac{1}{T} \int_0^T r(t, \delta t) dt. \tag{2.10}$$

In the case $\langle r(t, \delta t) \rangle = 0$ we thus obtain $\sigma_r(t, \delta t) = +\sqrt{r(t, \delta t)^2} = |r(t, \delta t)|$.

Figure 2.1 shows the variation of price $s(t)$, log-price return $r(t, 1 \text{ day})$, and absolute value of log-price returns $|r(t, 1 \text{ day})|$ for daily Dow Jones Industrial Average (DJIA) data over the period 2000 to mid-2011. An important feature of financial data, and clearly visible in Figure 2.1(c), is the occurrence of clustered regions where the volatility is much greater than the average value. We will return to this so-called *stylized fact* of financial data on various occasions throughout the book. It is not captured in the simple Brownian motion model of stocks, as discussed in Chapter 6.

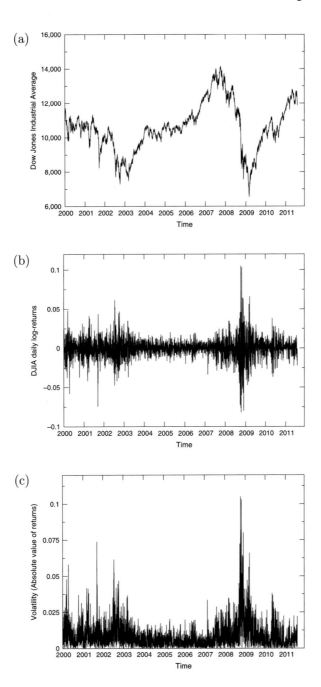

Figure 2.1 Daily data for the Dow Jones Industrial Average (DJIA) over the period 2000 to mid-2011: (a) price $s(t)$; (b) log-price return $r(t, 1\ \text{day})$; (c) absolute value of log-price returns $|r(t, 1\ \text{day})|$ (instantaneous volatility).

2.1.3 Quantifying temporal correlations

A measure of how the value of a variable x at time t is influencing its value at time $t + \tau$ is given by the evaluation of the *autocorrelation function*, $G(\tau)$, defined as

$$G(\tau) = \langle x(t)x(t + \tau) \rangle = \lim_{T \to \infty} \frac{1}{T} \int_0^T x(t)x(t + \tau)dt. \tag{2.11}$$

In the case of a set of N discrete values, such as prices or returns recorded at times t_i, the RHS is approximated by $\frac{1}{N}\sum_i x(t_i)x(t_i + \tau)$. (Applied to values of log-price return we may write $G(\tau, \delta t) = \langle r(t, \delta t)r(t + \tau, \delta t)\rangle$.)

Provided the standard deviation, $\sigma_t = \sqrt{\langle x^2(t)\rangle - \langle x(t)\rangle^2}$ is non-zero and finite, it is also common to define a *normalized autocorrelation function* $R(\tau)$ by

$$R(\tau) = \langle (x(t) - \langle x(t)\rangle)(x(t + \tau) - \langle x(t + \tau)\rangle)\rangle/(\sigma_t\sigma_{t+\tau}) = (G(\tau) - \langle x\rangle^2)/(\sigma_t\sigma_{t+\tau}). \tag{2.12}$$

The values of R are in the interval [-1:1], with 1 indicating perfect correlation, -1 perfect anti-correlation, and 0 no correlation.

The *autocovariance function* $C_{XX}(\tau)$ is defined by

$$C_{XX}(\tau) = \langle (x(t) - \langle x(t)\rangle)(x(t + \tau) - \langle x(t + \tau)\rangle)\rangle = \langle x(t)x(t + \tau)\rangle - \langle x(t)\rangle\langle x(t + \tau)\rangle, \tag{2.13}$$

i.e. $C_{XX}(\tau) = R(\tau)\sigma_t\sigma_{t+\tau}$.

For a stationary process, i.e. $\langle x(t)\rangle = \langle x(t + \tau)\rangle$ and $\sigma_t = \sigma_{t+\tau} = \sigma$, this reduces to

$$C_{XX}(\tau) = G(\tau) - \langle x(t)\rangle^2 = \sigma^2 R(\tau). \tag{2.14}$$

In order to evaluate correlations between *different* time series, e.g. correlations between different stocks, we will in Section 3.1.4 introduce a covariance matrix (see also Section 11.3).

For so-called *ergodic systems* a time average is equal to an ensemble average, see Section 1.4. We have in Eqns (2.11) and (2.14) already implicitly made this assumption when interchanging the definition of a temporal average (i.e. an integral over time) with the notation $\langle...\rangle$ which generally refers to an ensemble average.

In Figure 2.2 we show the (normalized) autocorrelation function $R(\tau, \delta t)$, eqn (2.12), for the log returns $r(t, \delta t)$ of the daily Dow Jones Industrial Average data of Figure 2.1. The function decays to zero essentially at the first time step and the daily returns appear to be independent. This is very similar to the case of having returns that are Gaussian distributed, see Figure 6.7, as is the case in the simple Brownian motion model of stocks which we will describe in Section 6.1.

However, the behaviour of the normalized autocorrelation function for the *volatility* is quite different. As a result of the clustered volatility that is evident in Figure 2.1(c), this function, shown in Figure 2.3, exhibits a very slow decay over quite a long time period of the order of a few hundred days. Clearly, the volatility fluctuations are *not* independent. Various research teams have suggested a power law decay with exponents between 0.2 and 0.3 over the first 100 days (Dacorogna *et al.*, 1993, Liu *et al.*, 1999, Roman *et al.* 2008, Shen and Zheng, 2009).

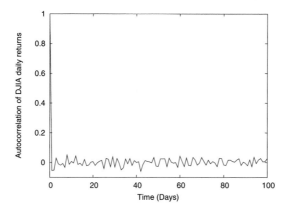

Figure 2.2 Plot of the normalized autocorrelation function $R(\tau, 1 \text{ day})$ of daily returns $r(t, 1 \text{ day})$ using Dow Jones Industrial Average data from 1993 to 2012. It is clear that there are no correlations certainly after the second day.

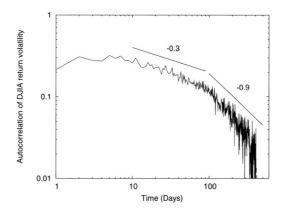

Figure 2.3 Auto-correlation $R(\tau, 1 \text{ day})$ of the *magnitude* (volatility) of daily DJIA log returns $|r(t, 1 \text{ day})|$ (data from 1993–2012), plotted on a log-log scale. Note that unlike the autocorrelation of the returns, here we see a very slow decay. Power law regimes may be identified for the ranges of 10–100 days, and exceeding 100 days, with approximate exponents of -0.3 and -0.9 (shown as lines of slope -0.3 and -0.9, respectively, in this double-logarithmic plot). We will return to a detailed analysis of this data in Sections 10.4 and 10.5.

We will in the following chapters detail the 'econophysics approach' of understanding the characteristic features of financial data, such as clustered volatility and slowly decaying autocorrelation for volatility. However, first we shall briefly describe some ways in which investors might assess financial data.

2.2 Two types of investors

2.2.1 Fundamental investors

Macro-economic data, company reports, and periodic attendance at company annual general meetings are the main concern of so-called fundamental investors who ask questions such as: Is the economy growing or shrinking? Are we heading for a recession or a period of strong economic activity? Is the company or the industrial sector including the company likely to follow a different path from the economy as a whole? What profits has the firm made over the past few years? What might the future trading prospects of a specific company be? Based on the profit forecast, what might the future earnings be?

An important measure is the price to earnings ratio, *P/E*. It is the ratio of share price to the earnings per share, *EPS*. By comparing P/E ratios for different companies one can get a feel for the relative value one has to pay for different stocks. Using historic values for earnings, one can compute historic P/E ratios. Using estimates (which are usually guesses) for future earnings, one can estimate future P/E ratios. If the company pays a dividend to shareholders, one can similarly calculate the yield which is the ratio of the dividend per share paid out to the share price.

On the basis of expected earnings, dividends, and P/E, one can get a feeling for whether the shares are 'cheap' or 'expensive'. By looking closely at other more intricate details, such as the value of sales relative to the value of the company, the value of debt to the company value, etc., a fundamental investor or accountant might come to a view as to whether the company is sound from a financial viewpoint.

In Table 2.1 we show a typical summary of the most important financial data for the company 'Restaurant Group', published in April 2013. Such tables are published regularly in the *Financial Times* or *Investors Chronicle*. They are also to be found on websites such as *www.digitallook.com*.

Examining trends in this kind of data, listening to expert opinion on future trends, and meeting with company executives sets the basis for decision-making in the mind of a *fundamental* investor.

Many fundamental investors become well-known for their views. One of the most famous fundamental investors is Warren Buffett, an American who until the emergence of Bill Gates and Microsoft was the wealthiest man in the world. Over the years, Buffett made money not just for himself, but also for many others who invested in his mutual fund. Widely known for his whimsical and folksy style of writing, one of the more endearing quotes contained in a recent annual report to his shareholders was: 'There seems to be some perverse human characteristic that makes easy things difficult'.

2.2.2 Technical traders

A chartist, noise, or technical trader relies on the price history and trends of the share price, rather than economic fundamentals. Indeed some chartists choose to ignore economic fundamentals completely. Chartists take the view that by following the path of share prices via the charts, future prices may be predicted.

The literature of the chartist is full of colourful language. One term that will be familiar to the physicist is 'momentum'. For chartists, this is a measure of the rate

Table 2.1 An example of financial data for the Restaurant Group (published in April 2013) that aims to inform shareholders and potential investors of the financial state of a company. Such data can be found from a variety of sources, such as *Investors Chronicle*, www.digitallook.com, finance.yahoo.com, www.google.com/finance. The core sources are the annually published accounts of the company.

<table>
<tr><td colspan="5" align="center">Restaurant Group</td></tr>
<tr><td colspan="5">Price: 481.60p, Market value: £964.47m</td></tr>
<tr><td colspan="5">12 month High: 488.10p Low: 269.20p</td></tr>
<tr><td colspan="5">Dividend yield: 2.4%, P/E ratio: 20</td></tr>
</table>

Year to 31 Dec	Turnover (£M)	Pre-tax profit (£M)	Earnings per share (pence)	Dividend per share (pence)
2007	367	42.8	14.9	7.3
2008	417	47.1	16.4	7.7
2009	436	48.3	18.9	8.0
2010	466	56.5	20.2	9.0
2011	487	48.6	21.9	10.5
2012	533	64.5	24.0	11.8
2013*	580	70.8	26.5	13.0

Normal market size 3000
⋆ Analysts estimate

of change of price with time. However, unlike the familiar units used in mechanics, there is no consensus as to the units to be used in finance. Some definitions appear to omit the time dimension completely. More interesting terms describe various patterns that occur in price time series. Interpolation lines are used to infer levels of 'relative strength' and 'support and resistance' levels. Chartists will also talk, for example, of 'head and shoulders, necklines, spikes, and islands'. Price formations where the price reaches a maximum, falls, and then rises before falling again are referred to as double tops. Triple tops may also occur. These are used to predict sell signals; equally 'bottoms, single, double, and triple' relate to possible buy signals. Basically the approach is to try to make forward projections using some kind of fit to a trend that can be seen in the empirical price data. The reader might be amused to learn that an article was published in the *Economist* magazine (19 August 2000, page 76) titled '*Heads, Shoulders and Broadening Bottoms*'. If not familiar with this jargon, one might think this was a reference to a type of hair shampoo!

In Figure 2.4, the bold jerky line in the upper graph is the share price and three moving averages (20, 50, and 200 days) are also plotted. The shaded region is defined by so-called *Bollinger bands* that seek to estimate the channel within which the fluctuations of the price are confined. Physicists familiar with polymers may think of the analogy with the channel swept out by a polymer as it reptates within a polymer network. The central graph shows volume of shares traded each day. (Light grey is the number bought; dark grey corresponds to sales). The lower graph is the so-called relative strength indicator, a momentum oscillator, computed by taking the net movements over a twenty day period. It is normalized such that it varies from 0 to 100.

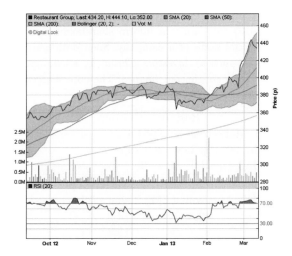

Figure 2.4 The evolution of the share price of Restaurant Group over the six months October 2012 to March 2013. For details see text.
(Figure reproduced with permission from www.digitallook.com.)

Some traders use this kind of indicator to generate buy and sell signals. So if it crosses the 30% value from below one might buy; if it crosses the 80% value from above one might sell.

The relationship between the chartist and the fundamentalist has not always been an easy one. Here is one quote from a well-known fundamental investor, Jim Slater. In 1992 he said: 'About 20 years ago I observed that chartists usually had dirty raincoats and large overdrafts. However, since those early days, I have met one or two who have made their fortune.'

More recently, one can see more recognition of the approach of the chartist from the fundamentalist camp. The latter will sometimes say things such as: 'Buy on weakness.' — 'Buy X because it is 20% undervalued compared with Y & Z.' — 'This gap should be closed.' — 'Is the rating out of line with historic norms?'. All these statements assume some assessment of the actual price, relative to some subjective norm which could be different for each investor.

The approach that we develop in the following chapters to analyse stock market movements differs radically from both that of technical and fundamental traders. We view markets as complex systems which are best understood in terms of the modelling of random processes. The appropriate methodologies have been developed and tested in the field of statistical physics. They date back as far as the eighteenth century with the work of Daniel Bernoulli, who gave a microscopic interpretation of Boyle's law for an ideal gas, with many important contributions in both nineteenth (Clausius, Maxwell, Boltzmann) and twentieth century (Einstein, Langevin, etc.).

We will start our discussion of the main ideas and their application to the analysis of financial data by introducing some basic concepts of probability theory.

3
Basics of probability

The laws of probability, so true in general, so fallacious in particular.
  Edward Gibbon (1737–94), English Historian.

The world around us encompasses a large variety of random systems. There are fluctuations everywhere. The movement of leaves on a tree, the motion of clouds in the sky, and the movement of traffic on a highway all fluctuate to a large or small degree. Figure 3.1 shows the outcome of a sequence of throws of a six-sided die, the simulated motion of a Brownian particle, and the temporal fluctuations of the Coca Cola stock price as yet further examples of random fluctuations.

In principle, one might think that everything could be calculated from first principles if only we knew the initial conditions, the nature of the dynamical laws, and the values of the parameters within these equations. For example, one might imagine from knowledge of the initial linear and angular velocity of the die, one could predict the trajectory and its subsequent motion so ultimately predict its final position. Doing such a calculation is, however, hopeless. Tiny errors we make would rapidly grow and our solution would become erroneous. Even if that were not to happen, our calculation would offer little insight beyond reproducing the specific outcome of one particular throw of the die. Fortunately, random (or stochastic) systems can be analysed in a different manner, offering more general information. The key is to calculate sensible quantities. For example, guessing the outcome of the throw of a die is no more than that, a guess. What is sensible to ask is, what is the *likelihood* or *probability* that the outcome of the throw will be, say, a five? Another good question might be what is the likelihood of observing a particular sequence of numbers?

3.1 Random variables

The central concept in the analysis of all stochastic systems is the random variable. This could be the outcome of the throw of a die or the position of a dart when thrown in a dartboard. The position of a molecule in a gas as it moves and collides with other molecules generates a path in space, $\mathbf{x}(t)$, that has the characteristics of a random walk. To a first approximation, the price of a financial asset as a function of time also has the appearance of a random walk and this analogy has been comprehensively exploited in the development of theories of quantitative finance.

While it is not possible to compute the particular outcome of a stochastic event (i.e. the result of a particular throw of a die), it may be possible to *compute the probability* that the random variable (in this case the top face of the die) takes on a certain value. In thinking about this, it is useful to imagine an ensemble of identical stochastic

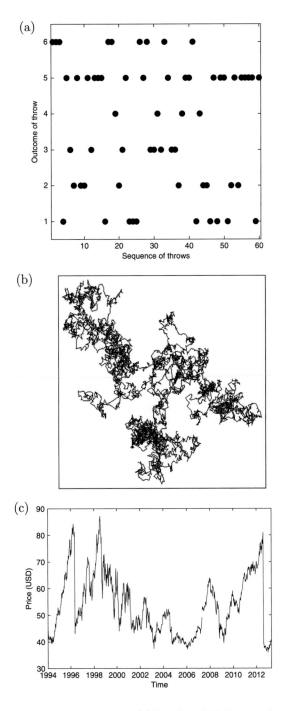

Figure 3.1 Examples of random behaviour: (a) Results of 60 throws of a six-sided die. (Data provided by Sam and Aran Hutzler.) (b) Computer simulation of the position of a Brownian particle moving in two dimensions. (c) Fluctuations of the Coca-Cola stock price in the period January 1994 to April 2013. (Weekly opening prices, data from Yahoo Finance.)

systems (e.g. a set of 100 identical dice). These will generate different values for the random variables, but each member of the ensemble will obey the same probability laws.

It is important to distinguish between the random variables that appear as the arguments of a distribution function, and particular values of the random variable that are the outcome of particular events.

A stochastic variable X takes on different values at random. Each value is associated with a certain probability p where $p \geq 0$. The possible values of X may be discrete, x_j, as in the case of the throw of a die, with associated probabilities $p(x_j)$, or continuous. In the latter case, say the measured weight of a die in grams, $p(x)dx$ is the probability that the value of X lies in the interval $[x, x + dx]$, and $p(x)$ is called the *probability density function*. The probabilities are normalized, i.e. $\sum_{i=1}^{N} p(x_i) = 1$ for the discrete case, where X can only take on values from the set $x_1, ..., x_N$ and

$$\int p(x)dx = 1, \tag{3.1}$$

for the continuous case, where the integral is over the entire continuous range of possible values of X. The probability density distribution may generally vary with time t, so we shall denote it as $p(x, t)$.

3.1.1   Average values and moments

We may now ask, for example, what is the probability that the value x, with $-\infty < x < \infty$), takes on values less or equal than some maximum value, X? It is given by the *cumulative* distribution function $C_<(X)$, defined by

$$C_<(X) = \int_{-\infty}^{X} p(x; t)dx. \tag{3.2}$$

We can define a *complementary* cumulative distribution function $C_>(X)$

$$C_>(X) = \int_{X}^{\infty} p(x, t)dx. \tag{3.3}$$

Evidently $C_<(X) + C_>(X) = 1$ as a consequence of the normalization of $p(x, t)$.

A value sometimes used to characterise the average of such distribution functions is the value of the *median* $x_{1/2}$ so that

$$C_>(x_{1/2}) = C_<(x_{1/2}) = 1/2. \tag{3.4}$$

Easy to define, this quantity is, however, not always easy to compute.

A second definition of an average value is the *most probable value*, x_{max}. This is defined as the value of x for which the derivative of $p(x, t)$ with respect to the variable x vanishes,

$$\frac{\partial p(x,t)}{\partial x}\bigg|_{x=x_{\max}} = 0. \tag{3.5}$$

Should this condition yield more than one solution, as might be the case when the distribution function is *multi-modal*, i.e. has more than one peak or maximum value, then the usual approach is to choose the value of x for which $p(x)$ is largest.

However, the mean value most frequently used because of its mathematical simplicity is the *first moment* m_1 of the probability distribution function, defined as follows[1]:

$$m_1 = m = \langle x(t) \rangle = \int_{-\infty}^{\infty} dx\, x\, p(x,t). \tag{3.6}$$

Unless the distribution function is both symmetric and uni-modal with a single peak value, all these various definitions for the average may differ. The differences may be important for the assessment of empirical distributions, but we shall, in what follows, for reasons of mathematical simplicity, limit ourselves to the definition given by eqn (3.6). Other definitions of the mean value are rarely used in the literature.

Higher order moments are readily defined. Thus the *second moment* is

$$m_2 = \langle x^2 \rangle = \int_{-\infty}^{\infty} dx\, x^2 p(x,t), \tag{3.7}$$

and the *n-th moments* are given by

$$m_n = \langle x^n \rangle = \int_{-\infty}^{\infty} dx\, x^n p(x,t). \tag{3.8}$$

Note that these moments need not always exist. For example, if the distribution function exhibits power-law tails for extreme values of the argument, x, then depending on the exponent associated with the power-law decay, the *standard deviation* σ, defined as

$$\sigma = \sqrt{\langle (x - \langle x \rangle)^2 \rangle} = \sqrt{\langle x^2 \rangle - \langle x \rangle^2}, \tag{3.9}$$

and higher order moments may diverge.

For physical systems, σ is intimately linked to a *diffusion constant*, as will be discussed in Sections 3.2 and 5.3. In the financial context it is often referred to as *volatility*, as was already mentioned in Section 2.1. The squared quantity σ^2 is called the *variance*.

All the moments are well defined and finite for the *Gaussian* or so-called *normal* distribution

$$p(x) = \frac{e^{-\frac{(x-m)^2}{2\sigma^2}}}{\sigma\sqrt{2\pi}} \tag{3.10}$$

with mean/first moment m and standard deviation σ.

---

[1]Economists usually talk about 'expected' values instead of 'mean' values and use the notation $E[x]$ rather than $m_1 = m = \langle x(t) \rangle$.

Exercise 3.1 The reader might wish to show that the symbol σ in eqn (3.10) is indeed the standard deviation, as defined by eqn (3.9).

However, for the *Cauchy-Lorentz distribution*, given by

$$p(x) = \frac{1}{\pi} \frac{\gamma}{(x-m)^2 + \gamma^2},\qquad (3.11)$$

only the most probable value and the median are finite, and coincide with the location of the maximum, m. None of the higher order moments, $n \geq 1$, exist. The half-width of the distribution at half-maximum is given by γ.

3.1.2 Moments and characteristic functions

Consider now the multivariate distribution, $p(\mathbf{x}, t)$ where the vector $\mathbf{x}$ is given by $\mathbf{x} = (x_1, \ldots, x_\alpha, \ldots, x_N)$ and t denotes time. The Fourier transform of the distribution, $\tilde{p}(\mathbf{k}, t)$, referred to as the *characteristic function*, can be readily seen to be the average of the exponential function,

$$\tilde{p}(\mathbf{k}, t) = \int e^{i\mathbf{k}\cdot\mathbf{x}} p(\mathbf{x}, t)\mathbf{dx} = \left\langle e^{i\mathbf{k}\cdot\mathbf{x}} \right\rangle_p. \qquad (3.12)$$

with $\mathbf{k} = (k_1, \ldots, k_\alpha, \ldots, k_N)$. From the normalization condition for $p(\mathbf{x}, t)$, we have $\tilde{p}(\mathbf{0}, t) = 1$. (Note that i is the imaginary unit, with $i^2 = -1$.)

The inverse Fourier transform is

$$p(\mathbf{x}; t) = \frac{1}{2\pi} \int e^{-i\mathbf{k}\cdot\mathbf{x}} \tilde{p}(\mathbf{k}; t)d\mathbf{k}, \qquad (3.13)$$

showing clearly that different characteristic functions yield different distribution functions. For simplicity in notation, in the rest of this section we shall drop the dependence on time and shall simply write $p(\mathbf{x})$ and $\tilde{p}(\mathbf{k})$.

All the moments of the distribution density $p(\boldsymbol{x})$ can be simply obtained from the characteristic function, eqn (3.12), by differentiation. Thus the first moment is given by

$$m_{1;\alpha} = \int d\mathbf{x} x_\alpha p(\mathbf{x}) = (-i) \left. \frac{\partial \tilde{p}(\mathbf{k})}{\partial k_\alpha} \right|_{\mathbf{k}=0}, \qquad (3.14)$$

and the second moment by

$$m_{2;\alpha_1,\alpha_2} = \int d\mathbf{x} x_{\alpha_1} x_{\alpha_2} p(\mathbf{x}) = (-i)^2 \left. \frac{\partial^2 \tilde{p}(\mathbf{k})}{\partial k_{\alpha_1} \partial k_{\alpha_2}} \right|_{\mathbf{k}=0}. \qquad (3.15)$$

In general we have

$$m_{n;\alpha_1\ldots\alpha_n} = (-i)^n \prod_{j=1}^{n} \left(\frac{\partial}{\partial k_{\alpha_j}} \right) \tilde{p}(\mathbf{k}) \, |_{\mathbf{k}=0}. \qquad (3.16)$$

Alternatively, if these moments are known, then the characteristic function may be constructed from its Taylor series expansion, as follows:

$$\tilde{p}(\mathbf{k}) = \sum_{n=0}^{\infty} \sum_{\{\alpha_1..\alpha_N\}} \frac{i^n}{n!} m_{n;\alpha_1,...\alpha_n} \left(\prod_{j=1}^{n} k_{\alpha_j} \right). \tag{3.17}$$

Exercise 3.2 The reader might wish to demonstrate eqn (3.17) for the one-dimensional case, $N = 1$.

For the Gaussian distribution of eqn (3.10), with mean zero, the characteristic function for the one dimensional case is readily shown to be

$$\tilde{p}(k) = \int_{-\infty}^{\infty} e^{ikx} p(x) dx = \int_{-\infty}^{\infty} dx e^{ikx} \frac{e^{-\frac{x^2}{2\sigma^2}}}{\sigma\sqrt{2\pi}}$$

$$= \frac{1}{\sigma\sqrt{2\pi}} \int_{-\infty}^{\infty} dx e^{-\frac{1}{2\sigma^2}[x-ik\sigma^2]^2} e^{\frac{-k^2\sigma^2}{2}}$$

$$= e^{-\frac{\sigma^2 k^2}{2}}. \tag{3.18}$$

We thus see that the characteristic function of a Gaussian is another Gaussian! The *moment-generating function* $M(\theta)$ is defined as

$$M(\theta) = \int e^{\theta x} p(x) dx = \langle e^{\theta x} \rangle. \tag{3.19}$$

Taylor expansion of the exponential immediately leads to $M(\theta) = 1 + \sum_{n=1}^{\infty} m_n \theta^n / n!$ where the m_n are the n-th order moments about zero. Note that unlike characteristic functions, moment-generating functions do not always exist.

3.1.3 Cumulants

Further useful functions are the cumulants. These are defined by the *cumulant generating function*, i.e. the logarithm of the characteristic function,

$$\Phi(\mathbf{k}, t) = \ln \tilde{p}(\mathbf{k}, t). \tag{3.20}$$

The *cumulants* are then defined as

$$c_{n;\alpha_1....\alpha_n} = (-i)^n \prod_{i=1}^{n} \left(\frac{\partial}{\partial k_{\alpha_i}} \right) \Phi(\mathbf{k}; t)|_{k_{\alpha_1}, k_{\alpha_2}, ...=0}. \tag{3.21}$$

Each cumulant of order n is a combination of the moments of order $l < n$.

$$c_{1;\alpha} = m_{1;\alpha}$$

$$c_{2;\alpha\beta} = m_{2;\alpha\beta} - m_{1;\alpha} m_{1;\beta} \tag{3.22}$$

$$c_{3;\alpha\beta\gamma} = m_{3;\alpha\beta\gamma} - m_{2;\alpha\beta} m_{1;\gamma} - m_{2;\beta\gamma} m_{1;\alpha} - m_{2;\gamma\alpha} m_{1;\beta} + 2m_{1;\alpha} m_{1;\beta} m_{1;\gamma}.$$

For a system with only one degree of freedom the cumulants reduce to

$$c_1 = m_1$$
$$c_2 = m_2 - m_1^2$$
$$c_3 = m_3 - 3m_2m_1 + 2m_1^3. \tag{3.23}$$

Exercise 3.3 The reader might like to prove the above relationships.

3.1.4 Further measures of probability functions

The second-order cumulant or *covariance matrix* $\mathbf{C}_{\alpha\beta}$ has, as its elements, the first order cumulants $c_{2;\alpha,\beta}$,

$$\mathbf{C}_{\alpha\beta} = \langle (x_\alpha - \langle x_\alpha \rangle)(x_\beta - \langle x_\beta \rangle) \rangle. \tag{3.24}$$

The covariance matrix is now seen to measure the degree to which values of x deviate, fluctuate or are dispersed about their mean values. For a system with one degree of freedom, it reduces to a single element:

$$c_2 = \langle (x - \langle x \rangle)^2 \rangle = m_2 - m_1^2 = \sigma^2. \tag{3.25}$$

In general we may introduce n-th order moments μ_n about the mean or *central moments*, where

$$\mu_n = \langle (x - m_1)^n \rangle. \tag{3.26}$$

Exercise 3.4 The reader may show the following relations.

$$c_1 = m_1$$
$$c_2 = \mu_2$$
$$c_3 = \mu_3$$
$$c_4 = \mu_4 - 3\mu_2^2. \tag{3.27}$$

It can be shown that either all but the first two cumulants vanish, or that there are an infinite number of them. In the first case the probability density is Gaussian.

Exercise 3.5 Show explicitly that for the Gaussian distribution defined by eqn (3.10) one obtains $\mu_4 = 3\mu_2^2$.

Two often used normalized measures indicating a deviation from a Gaussian are the *Skewness*, λ_3, and the *(excess) kurtosis*, λ_4, defined as,

$$\lambda_3 = \frac{c_3}{c_2^{3/2}} = \frac{\mu_3}{\sigma^3}$$

$$\lambda_4 = \frac{c_4}{c_2^2} = \frac{\mu_4}{\sigma^4} - 3. \tag{3.28}$$

While skewness measures the degree of asymmetry of the distribution, the excess kurtosis relates to the deviation of the tails of the distribution, as compared to a Gaussian; see Figure 3.2 for illustrating examples.

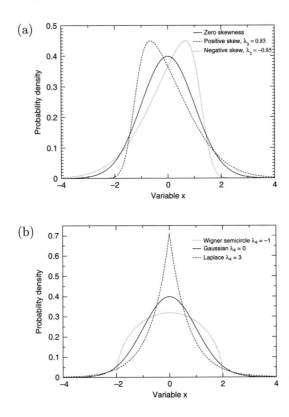

Figure 3.2 Illustrations of how the values of skewness and kurtosis determine the shape of distributions (the mean is set to zero and the variance to one in all examples). (a) Gaussian, together with two skewed Gaussians ($\lambda_3 = \pm 0.85$). (b) Gaussian, together with a Wigner semicircle distribution and a Laplace distribution.

Exercise 3.6 Show that for a Gaussian distribution, both skewness and kurtosis are zero.

3.2 Adding random variables

The distribution function $p(z)$ of the sum $z = x + y$ of two random variables x and y is given by

$$p(z) = \iint p(x, y)\delta(z - x - y)dxdy, \tag{3.29}$$

where $p(x, y)$ is the *joint probability density function* and $\delta(x\text{-}x_0)$ is Dirac's delta function, as defined in Appendix 3.4. The *joint probability* $p(x, y)dxdy$ gives the probability that x lies in the interval $[x, x + dx]$ and y in $[y, y + dy]$.

The characteristic function $\tilde{p}(k)$ for $p(z)$ is given by

$$\tilde{p}(k) = \langle\exp(ik[x+y])\rangle = \iint e^{ik(x+y)}p(x,y)dxdy. \tag{3.30}$$

If the variables x and y are independent, as is for example the case if they are generated by throwing two separate dice, then the joint probability distribution function is given by the product of two distribution functions $p_1(x)$ and $p_2(y)$ for x and y, respectively,

$$p(x,y) = p_1(x)p_2(y). \tag{3.31}$$

The integrals over x and y in eqn (3.30) then separate, leading to the following expression for $\tilde{p}(k)$:

$$\tilde{p}(k) = \tilde{p}_1(k)\tilde{p}_2(k). \tag{3.32}$$

We thus see that the characteristic function of a sum of two independent random variables is the product of the characteristic functions of the separate variables. In terms of the cumulant generating function, eqn (3.20), this may be written as

$$\Phi_z(k) = \Phi_x(k) + \Phi_y(k). \tag{3.33}$$

The above argument can be readily extended to the case where z is the sum of N variables,

$$z = \sum_{i=1}^{N} x_i. \tag{3.34}$$

The distribution function for z is then given by

$$p(z) = \frac{1}{N}\iint \cdots \int p(x_1,x_2,....,x_N)\delta(z - \sum_{i=1}^{N} x_i)\,dx_1\cdots dx_N. \tag{3.35}$$

Now suppose again that the variables x_i are independent and all drawn from the same distribution, $p(x)$, i.e.

$$p(x_1,.....,x_N) = \prod_{i=1}^{N} p(x_i) \tag{3.36}$$

As in the previous case of the sum of two variables we can insert this product into the expression for the characteristic function for $p(z)$. Separation of the N integrals leads to the result that the characteristic function $\tilde{p}(k)$ may be written as a product of the individual characteristic functions for the set of x_i:

$$\tilde{p}(k) = \tilde{p}^N(k) \tag{3.37}$$

In the case where all moments of the distribution of the increments x are finite, and considering the limit $N \to \infty$, where we can neglect higher order terms in the

Taylor expansion of the exponentials, we can rewrite this as follows,

$$\tilde{p}(k) = \langle \exp ikx \rangle^N = \left[1 + ik\langle x \rangle - \frac{k^2\langle x^2 \rangle}{2} + O(k^3) \right]^N$$

$$= \left[\exp \left(ik\langle x \rangle - [\langle x^2 \rangle - \langle x \rangle^2]k^2/2 \right) \right]^N$$

$$= \exp(iNk\langle x \rangle - N\sigma^2 k^2/2). \tag{3.38}$$

Here, σ^2 is the variance of p(x), defined in eqn (3.9).

Inverting the Fourier transform (using the method previously employed to derive eqn (3.18)) yields

$$p(z) = \frac{1}{2\pi} \int_{-\infty}^{\infty} e^{-[ikz - ikN\langle x \rangle + k^2 N\sigma^2/2]} dk = \sqrt{\frac{1}{2\pi N\sigma^2}} e^{-\frac{(z - N<x>)^2}{2N\sigma^2}}. \tag{3.39}$$

The interesting feature of this result is that, given our approximations, we have shown that the distribution function for the sum of N independent variables converges to a Gaussian, regardless of the precise form for the distribution function, $p(x)$ of the individual independent variables. We only need to assume that all the moments are finite and that it is possible to neglect the higher order terms in the series expansion. This result is referred to as the *central limit theorem*.

We see that in the limit $N \to \infty$ the width of this Gaussian distribution scales according to the relation

$$\sigma^2(z) = N\sigma^2. \tag{3.40}$$

We also deduce that in this limit we have $\langle z \rangle \to N\langle x \rangle$. This result is usually termed the *law of large numbers*. As with the central limit theorem, it is independent of the form of the initial distribution.

The final result for the distribution $p(z)$ may now be expressed as

$$p(z) = \frac{e^{-\frac{(z - <z>)^2}{2\sigma^2(z)}}}{\sqrt{2\pi\sigma^2(z)}}. \tag{3.41}$$

If the initial distribution function $p(x)$ for the independent events that comprise the sum is a Gaussian, then so is the distribution function for the sum. The shape of the distribution function is preserved during the summation process. Such a distribution function is said to be a *stable* distribution. Lévy distributions form another set of stable distributions, but with infinite variance. We shall discuss these in Chapter 9, in the context of scaling laws for financial data.

It is convenient at this stage of our discussion to introduce the concept of a (one-dimensional) *random walk*. We may interpret eqn (3.34) as the distance from the origin, $z(t)$, of a particle which moves with constant speed, as it travels in a time t a sequence of N steps x_i, randomly chosen from a distribution $p(x)$.

The mean distance travelled after N steps will be $\langle z \rangle = N\langle x \rangle = v_d t$, where v_d is the so-called *drift velocity*. If we also introduce the *diffusion constant* D through the relation

$$2Dt = N\sigma^2 = \sigma^2(z),\tag{3.42}$$

(dimension [length^2 time^{-1}]), we obtain the following expression for $\rho(z)$ (see eqn (3.41)),

$$p(z,t) = \sqrt{\frac{1}{4\pi Dt}} e^{\frac{-(z-v_d t)^2}{4Dt}}.\tag{3.43}$$

This form of $p(z)$ is familiar to those who have studied diffusion and Brownian motion– a topic we discuss in more detail in Chapter 5.

Exercise 3.7 Show that eqn (3.43) is a solution of the partial differential equation

$$\frac{\partial p(z,t)}{\partial t} = D\frac{\partial^2 p(z,t)}{\partial z^2} - \frac{\partial}{\partial z}[v_d p(z,t)].\tag{3.44}$$

3.3 Bayes' theorem

3.3.1 Axiomatic probability theory

The Russian mathematician, Andrey Kolmogorov (1903–87), developed an axiomatic approach to probability theory using set theory that allows its concise formulation. The axioms are defined in an abstract manner by introducing a family of subsets of a set Ω. A non-negative number, called the probability, is assigned to each subset. The probability, $P(\Omega)$, assigned to the set Ω is unity. A simple illustration is provided by a six-sided die. The subsets are associated with the individual faces, labelled 1 to 6, and the set Ω is the complete set of six faces. If the die is unbiased then the natural choice would be to assign to each face, the probability $1/6$. It is important to point out here that there is no axiom that yields the values for the probabilities. The a priori values ascribed are dependent on a subjective assessment.

Exercise 3.8 Assuming a perfect die, what is the probability to obtain the number 4 only four times in a total of 60 throws, as was the case for the experimental data shown in Figure 3.1(a)? Compare this with the probability of obtaining the number six times.

If each member, A_i, of the family of subsets $\{A_1 \ldots A_i \ldots A_N\}$ is *disjoint*, i.e., $A_i \cap A_j = \emptyset$ for all $i \neq j$ (the A_i have no elements in common), then the probability of combined subsets is the sum of the probabilities, $P(\bigcup_i A_i) = \sum_i P(A_i)$.

Returning to the example of our die, A_1 could be the number '2' and A_2 could be the set of odd numbers. Then $P(A \cup B) = P('2') + P(\text{odd}) = 1/6 + 3/6 = 2/3$.

3.3.2 Conditional probability and Bayes' theorem

Thomas Bayes (1701–1761) was an eighteenth century clergyman. Like many educated men of his time, he was also an amateur scientist and mathematician. His contribution to probability theory arose out of an interest in answering questions linked to gambling, and also to underpin the new area of insurance. The specific problem that was solved by Bayes was how to compute an *inverse* probability. He solved this issue by introducing the important concept of *conditional probability*.

Example 3.1 Prior to Bayes, mathematicians knew how to find the probability that, say, two people aged 85 die in a given year out of a sample of 45, given that the probability of any one of them dying was known. Thus if the probability of any one of the sample population dying is p, the probability of not dying is $1 - p$. So the probability that only two out of the population of 45 die in the year is $p^2(1-p)^{43}$.

It was however not known how to find *inverse* or *conditional* probabilities, for example, what is the probability that one member of the population will die, given that two out of the 45 have already died? In this case, of course, the solution is simple. If two people have died we are left with only 43 people and the probability that only one person now dies is $p(1-p)^{42}$. Bayes' method, published after his death in 1783, provides such solutions for many more difficult kinds of problems. It is now used to study numerous complex systems such as are found in astrophysics, weather forecasting, and decision-making in industry, clinical decision-making, pinpointing the outbreak of epidemics, and criminal investigations.

Conditional probability concerns the probability of an event A, given that B also occurs. For such an event to occur, the intersection A∩B must be non-zero. The conditional probability is then defined as:

$$P(A|B) = P(A \cap B)/P(B). \tag{3.45}$$

When a set Ω can be constructed from a family of N non-intersecting subsets $\Omega = A_1 \cup A_2 \cup ..... \cup A_N$, then it follows that the probability of an arbitrary set, X is given by:

$$P(X) = P(X \cap \Omega) = P(X \cap \bigcup_i A_i)$$

$$= P(X \cap A_1) + P(X \cap A_2) + ...P(X \cap A_N). \tag{3.46}$$

Using the expression for the conditional probability, eqn (3.45), this may be rewritten as

$$P(X) = P(A_1)P(X|A_1) + P(A_2)P(X|A_2) + ... + P(A_N)P(X|A_N). \tag{3.47}$$

This expression for $P(X)$ is usually termed the *total probability theorem*. It follows from eqn (3.45) that

$$P(X)P(A_i|X) = P(A_i \cap X) = P(X \cap A_i) = P(A_i)P(X|A_i). \tag{3.48}$$

Substitution into eqn (3.47) yields

$$P(A_i|X) = \frac{P(A_i)P(X|A_i)}{P(A_1)P(X|A_1) + P(A_2)P(X|A_2) + ... + P(A_N)P(X|A_N)}. \tag{3.49}$$

This is the celebrated theorem of Bayes that provides the relation between the conditional probabilities $P(A_i|X)$ and $P(X|A_i)$. It is immediately evident that we are required to know the set of a priori probabilities $\{P(A_i)\}$. Yet the formal theory provides no help with this. This lack of any guidance has provoked much debate and discussion, leading to so-called *Bayesian* and *frequentist* schools of thought. Most physicists follow the line of reasoning offered by *Bayesians* where subjective judgment is used to allocate a priori probabilities.

Example 3.2 Returning to our simple example: the probability that two of our elderly population of 45 die in the year, $P(2)_{45} = p^2(1-p)^{43}$ where p is the probability that any one person will die in the year. We want, $P(1|2)$, the probability that one member of the population will die, given that two have already died. $P(1|2) = P(1)P(2|1)/P(2) = pP(2)_{44}/p^2 = pp^2(1-p)^{42}/p^2 = p(1-p)^{42}$. This we readily see is $P(1)_{43}$ as we expect.

Example 3.3 A slightly more complex example is provided by the following. Say we have a pile of a million coins and we know that ten have two heads. The a priori probability that any selected coin is two-headed is 10^{-5}. Suppose we now select a coin at random and toss it five times and five heads are obtained. What is the probability that this coin has two heads? Bayes' theorem provides the answer.

Event A is two-headed coin, event B is fair coin and event C is five heads in succession. Therefore $P(A) = 10/10^6 = 10^{-5}$ and $P(B) = 1 - P(A)$. Clearly $P(C|A) = 1$, whereas $P(C|B) = (1/2)^5 = 0.03125$.

The answer we require is $P(A|C)$ which we can compute using Bayes' theorem.

$$P(A|C) = \frac{P(C|A)P(A)}{P(C|A)P(A) + P(C|B)P(B)}$$

$$= \frac{10^{-5}}{10^{-5} + (1/2)^5(1 - 10^{-5})} \simeq 32 \times 10^{-5}$$

Our assessment of the probability of obtaining a double-headed coin has now changed as a result of new information. This new, updated *a posteriori* probability can only be deduced by using a priori probabilities in conjunction with Bayes' theorem.

Example 3.4 Analysing certain stock data you deduce that the probability that it is both August and that the stock falls is 0.04. What is the probability that the stock falls, given that it is August?

The probability that it is August is 1/12. Now use Bayes' theorem to obtain the relevant conditional probability.

$$P(\text{fall}|\text{August}) = \frac{P(\text{fall} \cap \text{August})}{P(\text{August})} = \frac{0.04}{1/12} = 0.48. \tag{3.50}$$

3.4   Appendix: The δ-function

Dirac's delta function, $\delta(x\text{-}x_0)$, is a *generalized function*, which may be defined by the following two properties.

1. It is zero for all values of $x \neq x_0$.
2. Under an integral, in combination with a continuous function $f(x)$, the following holds,

$$\int_a^b \delta(x - x_0)f(x) \, dx \; = \; f(x_0) \quad \text{for } a \leq x_0 \leq b. \tag{3.51}$$

For the case $f(x) = 1$ one thus obtains

$$\int_a^b \delta(x - x_0)dx \; = \; \begin{cases} 1 & \text{if } a \leq x_0 \leq b \\ 0 & \text{otherwise.} \end{cases} \tag{3.52}$$

Note that the formal definition of the δ-function is dealt with in the mathematical theory of *distributions*.

4

Time dependent processes and the Chapman-Kolmogorov equation

Each move is dictated by the previous one – that is the meaning of order. If we start being arbitrary it'll just be a shambles: at least, let us hope so. Because if we happened, just happened to discover, or even suspect, that our spontaneity was part of their order, we'd know that we were lost.

Tom Stoppard, 'Rosencrantz and Guildenstern are Dead'.
(©1964 by Tom Stoppard)

One way a physicist may proceed when looking at complex systems is via some kind of equation of motion in analogy to Newton's laws of simple particulate motion. If we imagine the price of an asset to be analogous to the position of a particle we might imagine that some kind of Newtonian dynamics could apply to this system. However, the change of the price of an asset is a *stochastic* or *random process*, and a better analogy is with the random or stochastic motion of a Brownian particle (random walk), as we will discuss in Chapter 6. Here, we shall examine the statistical dynamics associated with such random motion. We shall see that the process involves a hierarchy of equations that in general are intractable. However, invoking the particular assumption that the process is *Markovian*, i.e. has a very short–term memory, it is possible to reduce this hierarchy down to a simple closed form, represented by the Chapman-Kolmogorov integral equation.

4.1 Multi-time stochastic processes

Suppose now, for a particular system, x is a dynamical variable of interest. It may be the position or velocity of a particle, a magnetisation vector, a stock price, the return of a stock price, or even the number of people in a large group holding a particular opinion. If the system were deterministic then we could, using the relevant equations of motion, determine $x(t)$ for each and every value of t, given that we know x at $t = 0$. But we are concerned with random variables for which such deterministic functions do not exist. What we can define, however, is the probability distribution density, $p(x,t)$, that at time t the dynamical variable, $x(t)$ takes the value x. More precisely, $p(x,t)\delta x$ is the probability that at time t the random variable x has a value between x and $x + \delta x$.

An intuitive idea of this random process is obtained by considering the evolution of the variable x with time. Figure 4.1 illustrates the idea. The black circles represent the most probable values of the variable x at times $t_1, t_2, \cdots, t_n$. The dashed line is the trajectory corresponding to probability maxima. It is clear that knowledge of

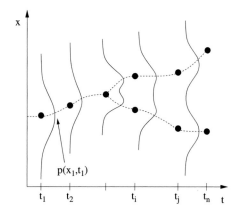

Figure 4.1 The evolution of a stochastic variable x with time t is described by the evolution of its probability distribution density $p(x,t)$. The dashed lines mark the most probable values of $x(t)$.

the distribution density, $p(x,t)$ is now not enough to describe the entire process. For example, the probability of finding both x_i at time t_i and x_j at time t_j requires knowledge of the *joint probability density* $p_2(x_i, t_i; x_j, t_j)$, introduced at the beginning of Section 3.2. A complete characterisation of the process requires knowledge of the joint probability density, $p_n(x_1,t_1; x_2,t_2;..; x_n, t_n)$ for all times $t_1, t_2, \cdots, t_n$.

In Section 3.1.2 we introduced multivariate distributions. When the many variables in such a distribution are related to each other via evolution of the random variable $x_j = x(t = t_j)$, the distribution refers to a stochastic process.

Since we are dealing with probabilities, the distributions $p_n(x_1,t_1; x_2,t_2;..; x_n, t_n)$ fulfil $p_n \geq 0$. Furthermore, higher order joint distributions must be compatible with lower-order joint distributions. For example

$$\int p(x_1, t_1; x_2, t_2; x_3, t_3)dx_3 = p(x_1, t_1; x_2, t_2);$$

$$\int p(x_1, t_1; x_2, t_2)dx_1 = p(x_2, t_2). \tag{4.1}$$

We may therefore think in terms of a hierarchy of functions extending ad infinitum. In order to compute these functions, p_n, it is usual to seek a route that allows higher order functions to be expressed in terms of the lower order ones. Perturbation methods or some kind of truncation process are the normal methods used. Before we discuss one such truncation process, we introduce a *stationary stochastic process* as a stochastic process for which the distribution function is independent of the origin of time:

$$p(x_1, t_1 + \tau; x_2, t_2 + \tau; ......x_n, t_n + \tau) = p(x_1, t_1; x_2, t_2; ......x_n, t_n)$$
$$\text{for all values of } \tau. \tag{4.2}$$

It immediately follows that, for a stationary process, the one-time distribution function is independent of time,

$$p(x,t) = p(x). \tag{4.3}$$

Similarly, the two-time joint probability distribution function obeys

$$p(x_2, t_2; x_1, t_1) = p(x_2, t_2 - t_1, x_1, 0). \tag{4.4}$$

4.2 Markov processes

One widely used truncation method assumes the stochastic process is Markovian. To simplify matters at this point we introduce a shorthand notation:

$$p(n; ...; 1) \equiv p_n(x_n, t_n; ...; x_1, t_1). \tag{4.5}$$

Furthermore we have arranged the indices so that they correspond to a time sequence $t_n > t_{n-1} > ... > t_2 > t_1$, which will be useful when working out conditional probabilities, see below. Finally, in physics one generally assumes that joint distributions are symmetric with respect to permutations of groups of variables with each other, i.e.

$$p_n(x_1, t_1; ...; x_i, t_i; ..; x_j, t_j; ...; x_n, t_n) = p_n(x_1, t_1; ...; x_j, t_j; ..; x_i, t_i; ..; x_n, t_n). \tag{4.6}$$

This property is, however, not necessary in order to deduce the expressions below.

For a random process where correlations are non-existent and the random fluctuations at different times are completely independent it follows that

$$p(n; ...; 2; 1) = p(n)....p(2)p(1). \tag{4.7}$$

This may be true if the time differences $t_2 - t_1$, $t_3 - t_2$, and so on are large. However, for sufficiently short times, we would expect correlations to be present as a result of causality and such an approximation to break down.

The next simplest approximation is the so-called *Markov process*. The idea of a Markov process is captured by imagining a person reading a book who, having a defective memory, can only recall the last sentence he or she has read, having forgotten all the previous sentences. Formally it is captured using a conditional probability, as introduced in Section 3.3.2. Specifically for the multivariate stochastic process introduced above, a Markov process corresponds to

$$p(n|n-1, ..., 2, 1) = p(n|n-1). \tag{4.8}$$

Using the definition of conditional probability and introducing the Markovian approximation yields:

$$p(n; n-1; ...; 1) = p(n|n-1; ...; 1)p(n-1; ...; 1) =$$
$$= p(n|n-1)p(n-1|n-2; ...; 1)p(n-2; ...; 1)$$
$$= p(n|n-1)p(n-1|n-2)p(n-2; ..; 1)$$
$$= \cdots = p(n|n-1)p(n-1|n-2).....p(2|1)p(1). \tag{4.9}$$

A Markov process is thus entirely characterized by a knowledge of the two-time conditional or so-called *transition probability distribution*, $p(i+1|i)$ and the initial distribution, $p(1)$.

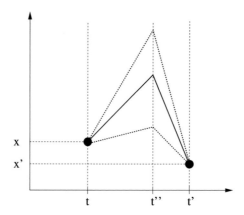

Figure 4.2 Pictorial representation of the Chapman-Kolmogorov equation. The integral in eqn (4.15) sums over all possible paths joining point (x, t) to point (x', t').

Integrating the transition probability over initial states i or final states f leads to

$$\int p(f|i)p(i)di = p(f),\qquad(4.10)$$

and

$$\int p(f|i)df = 1,\qquad(4.11)$$

respectively, where we have used the definition of the joint probability density, $p(f|i) = p(f;i)/p(i)$.

4.3 The Chapman-Kolmogorov equation

We will now derive the Chapman-Kolmogorov equation (see also Figure 4.2). First we note that, following from eqn (4.1), integration of the three-state joint probability $p(3; 2; 1)$ over the second variable results in

$$p(3;1) = \int p(3;2;1)d2,\qquad(4.12)$$

where we have used the shorthand notation $\int d2 \equiv \int dx_2$.

Substituting $p(3; 2; 1) = p(3|2)p(2|1)p(1)$ we obtain

$$p(3;1) = \int p(3|2)p(2|1)p(1)d2.\qquad(4.13)$$

Using the definition of the conditional probability distribution function for the term on the left hand side we finally obtain

$$p(3|1) = \int p(3|2)p(2|1)d2.\qquad(4.14)$$

This is the celebrated *Chapman-Kolmogorov equation* (sometimes called the *Einstein-Smoluchowski equation*). Returning to the full notation this is

$$p(x', t'|x, t) = \int p(x', t'|x'', t'')p(x'', t''|x, t)dx'',$$
(4.15)

where we have set $x = x_1$, $x'' = x_2$ and $x' = x_3$.

Let us return to our initial picture of the stochastic process as the evolution of a probability distribution, together with a line in a two-dimensional space that corresponds to the most probable value, see Figure 4.1. We can now imagine the transition probability to be constructed from the sum over all possible paths of the product of two such transitions that pass through the intermediate point, $x''(t'')$, as illustrated in Figure 4.2. (A similar pictorial approach was undertaken by Richard Feynman for path integral methods in quantum field theory.)

Note that an equation for the one time distribution function may be obtained by inserting $p(x', t'|x, t) = p(x', t'; x, t)/p(x, t)$ into the Kolmogorov eqn (4.15) and integrating over dx. This leads to

$$p(x', t') = \int p(x', t'|x'', t'')p(x'', t'')dx'',$$
(4.16)

as was already noted earlier, see eqn (4.10).

5

The Langevin approach to modelling Brownian motion

And when I'm done, I want you to do the following: look at those numbers, all those little numbers and Greek letters on the board, and repeat to yourselves, 'This is reality', repeat it over and over.

Anon

In 1828, the botanist Robert Brown (1773–1858) was examining small grains of pollen in water using a microscope and saw that they appeared to move, even though there was no apparent movement of the water. He concluded that the movement belonged to the grains themselves. Brown was not the first to make this discovery. His observations were preceded by those of two Dutchmen, the inventor of the microscope Anton van Leuwenhoek (1632–1723) and Jan Ingenhousz (1730–99). Numerous explanations of the phenomenon were proposed, and it was even speculated that it could be linked to the origin of life itself.

Most physicists know that Einstein put forward the explanation that the motion of the particles was due to impact with the individual molecules of water whose velocities followed the Maxwell-Boltzmann distribution (Einstein, 1905). It is not so commonly known within the physics community that the French researcher Bachelier developed independently the same statistical analysis and applied it to fluctuations of stocks five years earlier in 1900. Bachelier spent most of his career in a small provincial French college and his work, published only in his PhD thesis (Bachelier, 1900), remained largely unnoticed for half a century, despite having had the eminent mathematician and physicist Poincaré as an examiner. We will return to Bachelier's approach in Chapter 6.

In this chapter we develop the essential points of the theory of Brownian motion. We follow Langevin's approach, developed in 1908, and described by him as 'infinitely more simple' than Einstein's derivation from 1905 (see Gardiner, 2004).

5.1 Langevin equations

Consider a Brownian particle of mass M and radius a, immersed in a fluid with viscosity, η and composed of small molecules of mass, $m \ll M$. Using Newton's second law we can write the equation of motion for the velocity, v, of the large Brownian particle as

$$M\dot{v} = -\zeta v + F(t) + B(x,t). \tag{5.1}$$

The first term on the RHS is the force due to friction, experienced as a result of the motion. Assuming classical hydrodynamics it can be shown (see for example Acheson, 1990) that, for spherical particles of radius a, the friction constant, ζ, is given by $\zeta = 6\pi\eta a$ (Stokes drag). The second term represents the random forces due to collisions with the water molecules. The third term captures all other forces from, say, gravity, electrical forces, etc. In what follows we shall neglect this third term since it is not essential to our argument. We shall also simplify the notation and thus eqn (5.1) reduces to

$$\frac{dv}{dt} = -\lambda v + A(t), \tag{5.2}$$

where $\lambda = \zeta/M$ (with dimension of an inverse time) and $A(t) = F(t)/M$.

Collisions of the fluid molecules with the Brownian particle are many and irregular in both strength and direction. An immediate consequence is that the solution to our problem is not deterministic. Rather the solution is probabilistic and depends critically on the properties of the random force, $A(t)$.

Equation (5.2) is called a *Langevin equation*.[1] In a more general form it may take the form

$$M\frac{dv(t)}{dt} = -\frac{dV(x)}{dx} + F(t) \; (+ \text{ velocity dependent term}), \tag{5.3}$$

describing the movement of a particle of mass M in a potential $V(x)$ and subject also to a fluctuation force $F(t)$.

5.2 The velocity distribution of a Brownian particle

It is straightforward to integrate eqn (5.2) and obtain an expression for the velocity v:

$$v(t) = v_0 e^{-\lambda t} + e^{-\lambda t} \int_0^t dt' e^{\lambda t'} A(t'), \tag{5.4}$$

where we set $v_0 = v(t = 0)$.

At this point we must make some assumptions about the collisions and the random force.

Assume that over the time, t, there are very many collisions between the Brownian particle and the fluid molecules. It is reasonable to assume that the average of the random force t is zero:

$$\langle A(t) \rangle = 0 \tag{5.5}$$

Consider a Brownian particle of radius $a \sim 10^{-6} m$ and mass $M \sim 10^{-14}$kg, suspended in water with viscosity, $\eta \sim 10^{-3}$ Pa s, which results in $\lambda \sim 10^6 \text{s}^{-1}$. The frequency of the molecular collisions is much larger and of the order of $10^{12} - 10^{13}\text{s}^{-1}$. The autocorrelation function $G(t-t') = \langle A(t)A(t') \rangle$ can thus be expected to fall off in a time of approximately 10^{-12}s, which is much faster than the decay time λ^{-1}. Therefore we can approximate the autocorrelation function by a Dirac delta function (see Appendix 3.4):

$$\langle A(t)A(t') \rangle = G(t - t') \simeq \alpha\delta(t - t'), \tag{5.6}$$

where α is an as yet unspecified constant.

---

[1] For a comprehensive discussion of such equations and their applications see Coffey *et al.* 2004.

Let us compute the *ensemble average* $\langle v(t) \rangle_{v_0}$ of the velocity $v(t)$, as given by eqn (5.4). The ensemble average is taken using the probability distribution of velocities for the fluid molecules and the subscript v_0 reminds us that the average is conditional on the initial velocity of the Brownian particle being v_0. Thus using eqn (5.5) we have

$$\langle v(t) \rangle_{v_0} = v_0 e^{-\lambda t}. \tag{5.7}$$

For small times, the average or mean velocity is v_0. For large times, the mean velocity tends to zero.

The mean square velocity $\langle v^2 \rangle_{v_0}$ is also readily obtained. Squaring eqn (5.4) and then forming the ensemble average gives

$$\langle v^2(t) \rangle_{v_0} = v_0^2 e^{-2\lambda t} + e^{-2\lambda t} \int_0^t dt_1 \int_0^t dt_2 e^{\lambda(t_1+t_2)} \langle A(t_1)A(t_2) \rangle_{v_0}, \tag{5.8}$$

where we have taken into account eqn (5.5). The exponential terms can be taken out of the averaging process since they vary on a much longer time scale than the molecular fluctuations.

Taking into account the absence of correlations in our random force term as expressed in eqn (5.6), the above equation reduces further to

$$\langle v^2(t) \rangle_{v_0} = v_0^2 e^{-2\lambda t} + \frac{\alpha}{2\lambda}(1 - e^{-2\lambda t}). \tag{5.9}$$

For our random force as defined by eqns (5.5) and (5.6) we thus obtain the following expression for the variance of the velocity

$$\langle \Delta v^2(t) \rangle_{v_0} = \langle \langle v^2(t) \rangle - \langle v(t) \rangle^2 \rangle_{v_0} = \frac{\alpha}{2\lambda}(1 - e^{-2\lambda t}). \tag{5.10}$$

At time $t = 0$ the fluctuations are zero as expected. As $t \to \infty$ they tend to the finite value $\frac{\alpha}{2\lambda}$.

At this point, rather than computing this value from first principles, we invoke the equipartition theorem from kinetic theory and thermodynamics. This teaches us that for a system in equilibrium, $\langle v^2 \rangle$ is proportional to the temperature, T. For a one-dimensional system we have $\frac{1}{2}M\langle v^2 \rangle = \frac{1}{2}k_B T$, where k_B is the Boltzmann constant and M is the mass of the Brownian particle. Comparing this result with eqn (5.9) we obtain in the limit of $t \to \infty$:

$$\alpha = 2\lambda k_B T/M. \tag{5.11}$$

Since we know that in this limit the mean velocity is zero (see eqn (5.7)), we may now use the result given by eqn (5.10) to compute directly the probability density distribution for the velocities of the Brownian particle, invoking the central limit theorem (Section 3.2). Thus,

$$P(v, t|v_0, 0) = \frac{1}{\sqrt{2\pi \langle \Delta v^2(t) \rangle_{v_0}}} e^{-\frac{v^2(t) - v_0^2 e^{-2\lambda t}}{2\langle \Delta v^2(t) \rangle_{v_0}}}. \tag{5.12}$$

In the limit $t \to \infty$ this tends to the stationary or steady state solution,

$$P(v) = \frac{1}{\sqrt{\pi\alpha/\lambda}}e^{-\frac{v^2}{\alpha/\lambda}}. \tag{5.13}$$

Inserting for α (eqn (5.11)) results in the *one-dimensional* equilibrium Maxwell-Boltzmann velocity distribution

$$P(v) = \left(\frac{M}{2\pi k_B T}\right)^{1/2} \exp\left[-\frac{Mv^2}{2k_B T}\right]. \tag{5.14}$$

Note that in three dimensions the factor outside the exponential is given by $\left(\frac{M}{2\pi k_B T}\right)^{3/2}$.

5.3   Modelling the position of a Brownian particle

Retaining our assumption that fluctuations are Gaussian, it is relatively simple to compute the corresponding distribution function for the *position* of our Brownian particle. The first step is to integrate eqn (5.4) for the velocity, $v(t)$ to obtain the position, x,

$$x(t) - x_0 = \frac{v_0}{\lambda}(1 - e^{-\lambda t}) + \int_0^t dt' e^{-\lambda t'} \int_0^{t'} dt'' e^{\lambda t''} A(t''). \tag{5.15}$$

The second term on the RHS may be integrated by parts to obtain,

$$x(t) - x_0 = \frac{v_0}{\lambda}(1 - e^{-\lambda t}) + \frac{1}{\lambda}\int_0^t dt'[1 - e^{\lambda(t-t')}]A(t'). \tag{5.16}$$

Exercise 5.1 The reader may wish to show this by setting $u = \int_0^{t'} e^{\lambda t''} A(t'')dt''$ and $v = -e^{-\lambda t}/\lambda$ for use in the rule for partial integration, $\int uv' = uv| - \int u'v$.

Forming the ensemble average as before gives the average position,

$$\langle x(t) \rangle = x_0 + \frac{v_0}{\lambda}(1 - e^{-\lambda t}) \tag{5.17}$$

The fluctuations are readily obtained from eqns (5.15) and (5.17), and using once again the approximation for the autocorrelation, eqn (5.5) results in

$$\langle[x(t) - \langle x(t) \rangle]^2\rangle = \frac{1}{\lambda^2}\left(\int_0^t dt'[1 - e^{\lambda(t-t')}]A(t')\right)^2 = \frac{\alpha}{\lambda^2}\int_0^t ds[1 - e^{-\lambda s}]^2. \tag{5.18}$$

Exercise 5.2 The reader may wish to verify this expression.

The integral can be solved exactly, however, for our purposes it is of interest to examine the limits of small and large t. For small t we expand the exponential to obtain

$$\langle[x(t) - \langle x(t)\rangle]^2\rangle \simeq \alpha t^3/3. \tag{5.19}$$

We can similarly expand the expression for the mean position given by eqn (5.17) to obtain

$$\langle x(t)\rangle \simeq x_0 + v_0 t. \tag{5.20}$$

For small values of time, we see that the mean position increases linearly with time and fluctuations, being of higher order in time, can be neglected.

However, in the limit of $t \to \infty$ the result is qualitatively different. To leading order in t we obtain

$$\langle x(t)\rangle \simeq x_0 + \frac{v_0}{\lambda}, \tag{5.21}$$

and

$$\sigma^2(t) = \langle[x(t) - (x_0 + \frac{v_0}{\lambda})]^2\rangle \simeq \frac{\alpha}{\lambda^2}t, \tag{5.22}$$

where we have neglected the term $e^{-\lambda s}$ in eqn (5.18) for large times.

The mean position remains bounded, being shifted only by a constant factor, albeit one that depends on the initial velocity; however, now the fluctuations increase linearly with time. The process is equivalent to a diffusion process with diffusion constant

$$D = \alpha/(2\lambda^2) = \frac{\sigma^2(t)}{2t} = k_B T/\zeta, \tag{5.23}$$

with dimension [length^2 time^{-1}] (ζ is the friction constant in eqn (5.1)), see eqns (3.42) and (5.11)). The conditional probability density distribution, assuming as before, Gaussian statistics that only require the first and second moments of x, is given by

$$P(x, t|x_0, 0) = \frac{1}{\sqrt{2\pi\sigma^2(t)}} e^{-\frac{[x - x_0 - \frac{v_0}{\lambda}(1 - e^{-\lambda t})]^2}{2\sigma^2(t)}}. \tag{5.24}$$

The width (dispersion) of the distribution increases linearly with time, governed by the diffusion constant D.

5.4 Beyond Brownian motion

An alternative approach to modelling Brownian motion, based on the Chapman-Kolmogorov equation (see Section 4.3), was developed independently by Bachelier and Einstein. Bachelier's interest concerned the description of fluctuations of stock prices, and we will show the merits and shortcomings of a model based on Gaussian fluctuations in Chapter 6.

In practise, the correlations are more complex than the description used here. For example, the δ-function approximation used to simplify eqn (5.6) breaks down, leading not to Gaussian distributions but other distributions that exhibit long-range *fat tails*, as we will discuss in Chapter 9.

Many stochastic processes exist in nature where the volatility, $\sigma^2(t)$, varies as a fractional power of time. The hydrologist Harold Hurst was the first to characterize the rise and fall of water in the river Nile by the relation $\sigma^2(t) \propto t^{2H}$, where H is now called the *Hurst exponent* (see for example Nicolis and Nicolis, 2007). Hurst obtained a value of $H = 0.7$, differing from the value of $H = 0.5$ for the uncorrelated stationary fluctuations with finite variance that we dealt with in this chapter (see eqn (5.22). Since Hurst exponents are a measure of the long–term memory of a stochastic process, they are also used to characterize financial fluctuations and explore deviations from Gaussian behaviour.

6

The Brownian motion model of asset prices

Es ist möglich, daß die hier zu behandelnden Bewegungen mit der sogenannten Brownschen Molekularbewegung identisch sind; die mir erreichbaren Angaben über letztere sind jedoch so ungenau, daß ich mir hierüber kein Urteil bilden konnte.

It is possible that the motions to be discussed here are identical with the so-called Brownian molecular motion; however, the data available to me on the latter is so imprecise that I could not form a judgment.

Albert Einstein, in his paper on Brownian motion from 1905.

In 1900, Bachelier, then a young graduate student in Paris, had published in his PhD thesis entitled 'Théorie de la Spéculation' a complete analysis of futures and options, based on the ideas of random fluctuations of stock prices, superimposed on a constant drift (Bachelier, 1900[1]).

In his development of the corresponding mathematical framework he predated Einstein, who in 1905 showed that the thermal motion of molecules results in a random motion of suspended particles, observable by microscopy, and possibly identical to Brownian motion (Einstein, 1905).

Einstein's work was immediately recognized as being highly relevant, not only since it provided a final proof for the existence of atoms, but also for its treatment of random motion. This is quite different to the perception of Bachelier's results, which remained largely unknown in the scientific and economic communities until the 1960s.

There are a number of issues with Bachelier's original model, and some of these must have become clear to Bachelier himself when comparing his predictions to actual financial data. For a detailed discussion we refer to the reader to the book by Voit (2001).

Since Bachelier considered price changes as being Gaussian distributed, one obvious shortcoming is the possibility of negative prices. This may be avoided by considering the *log*-prices as being Gaussian distributed, resulting in the (geometric) Brownian motion model of stock prices which we will introduce below. (Bachelier's original model is at times referred to as *arithmetic* Brownian motion model of stock prices.)

This model is also at the heart of the work by Black, Scholes, and Merton on option prices from 1973 (see Chapter 8). Interesting to know is that this was despite the fact that it had been shown by Mandelbrot by that time that the distribution function

---

[1] See Davis and Etheridge, 2006, for a recent annotated translation into English, together with a brief history of stochastic analysis and financial economics since Bachelier.

for asset prices deviates significantly from the Gaussian distributions that feature in Bachelier's work. Even today, with our more complete knowledge of asset price fluctuations, Gaussian fluctuations are frequently used in the commercial arena! As a result, the risk involved in certain financial transactions is often grossly underestimated with, as we shall see, disastrous consequences.

In the following we will give a treatment of random motion based on the Chapman–Kolmogorov equation (see Section 4.3), similar to Bachelier's original work. The alternative treatment in terms of a Langevin equation was already given in Chapter 5. Both approaches have their merits, and will be in use throughout the following chapters.

6.1 Modelling the distribution of returns

Let us recall the Chapman–Kolmogorov equation for the conditional probability distribution function,

$$p(u, t|u_0, t_0) = \int p(u, t|u', t')p(u', t'|u_0, t_0)du'. \tag{6.1}$$

In the geometric Brownian motion model of stock prices the stochastic variable $u(t)$ is identified with the logarithm of the price, $u(t) = \ln s(t)$. It is then assumed that the conditional probability distribution depends only on the differences $u(t) - u(t_0), u(t') - u(t_0)$, etc. Fluctuations at different times are taken as independent of each other, drawn from the same distribution function (*i.i.d.* random variables). This implies that

$$p(u, t|u_0, t_0) = p(u - u_0, t - t_0|u_0, t_0) = p(u - u_0, t - t_0). \tag{6.2}$$

Substitution into (6.1) yields

$$p(u - u_0, t - t_0) = \int p(u - u', t - t')p(u' - u_0, t' - t_0)dx'. \tag{6.3}$$

Introducing the log-returns $r(t_0, \delta t) = u - u_0 = \ln s(t_0 + \delta t) - \ln s_0$ and $r'(t_0, \delta t') = u' - u_0 = \ln s(t_0 + \delta t') - \ln s(t_0)$ this is re-written as,

$$p(r(t_0, \delta t), t - t_0) = \int p(r - r', t - t')p(r', t' - t_0)dr'. \tag{6.4}$$

Using the Fourier convolution theorem we readily obtain

$$\tilde{p}(k, \delta t) = \tilde{p}(k, \delta t - \delta t')\tilde{p}(k, \delta t'). \tag{6.5}$$

The solution of eqn (6.5) may be expressed in the following manner,

$$\tilde{p}(k, \Delta t) = e^{iF(k)\Delta t}, \tag{6.6}$$

where the function $F(k)$ is for the moment an arbitrary function of k and independent of the time Δt. For any value of Δt the probability distribution $\tilde{p}(k, \Delta t)$ should be normalized and hence $\tilde{p}(0, \Delta t) = 1$. This can be ensured if we choose $F(0) = 0$.

Now we can, following the account in Section 3.1, obtain the various moments of the probability distribution function. Differentiating eqn (6.6) with respect to k and using eqns (3.14) and (3.15), respectively, gives values for the first and second moments,

$$m_1 = -i\frac{\partial \tilde{p}(k, \delta t)}{\partial k}\bigg|_{k=0} = F'(0)\delta t = v_d \delta t = \langle r(t, \delta t)\rangle, \tag{6.7}$$

$$m_2 = (-i)^2 \frac{\partial^2 \tilde{p}(k, \delta t)}{\partial k^2}\bigg|_{k=0} = -iF''(0)\delta t + \delta t^2 F'^2(0) = \langle r^2(t, \delta t)\rangle. \tag{6.8}$$

The parameter, $v_d = F'(0)$, i.e. the rate of increase of the average return, as a function of time interval δt, is constant.

Exercise 6.1 We shall leave the derivation of the above results as an exercise for the reader. (Hint: $F(0) = 0$.)

The volatility (or variance) $\sigma(\delta t)$ is then given by (see eqn (3.25)),

$$\sigma^2(\delta t) = \langle (r(t, \delta t) - \langle r(t, \delta t)\rangle)^2\rangle = m_2 - m_1^2 = -iF''(0)\delta t \equiv \sigma_0^2 \delta t \tag{6.9}$$

where we have renamed the constant $F''(0)$ as $\sigma_0^2 = F''(0)$.

Thus in this theory the random variable $r(t, \delta t)$, i.e. the log-price return, has both an average value and a variance that increase linearly in time δt (eqns (6.7) and (6.9), respectively).

Taylor expanding $F(k)$ in eqn (6.6) up to second order in k leads to

$$\tilde{p}(k, \delta t) = \exp\left[i\langle r(t, \delta t)\rangle k - \frac{\sigma_0^2 \delta t}{2}k^2\right]. \tag{6.10}$$

Performing an inverse Fourier transformation of this then gives the final result that the log-returns satisfy the Gaussian distribution,

$$p(r, \delta t) = \frac{1}{\sqrt{2\pi\sigma_0^2 \delta t}} \exp\left[-(r - v_d \delta t)^2/(2\sigma_0^2 \delta t)\right]. \tag{6.11}$$

This probability distribution of the log-returns x satisfies the diffusion equation for Brownian motion with a drift, as was introduced earlier, see eqn (3.44). The coefficient σ_0^2 is thus identified as proportional to the *diffusion constant* D, i.e. $\sigma_0^2 = 2D$ and the coefficient v_d with a *drift velocity* associated with the evolution of the probability distribution. Note that we reproduce the results for the motion of a Brownian particle, derived from a Langevin equation (see Section 5.3) if we set the drift velocity $v_d = 0$.

In Bachelier's time, the data available was for relatively low frequencies, typically weeks or even years. Figure 6.1 shows the cumulative distribution of log-price annual returns over the period 1800 to 2001, together with the cumulative distribution for a Gaussian. It seems that such a fit is fairly reasonable, although looking carefully at the data especially in the top half of the graph already one can see that the actual

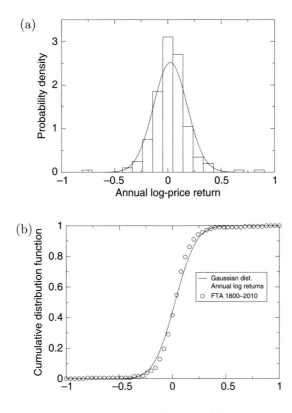

Figure 6.1 Distribution and cumulative distribution of log-price *annual* returns for the FTSE All-Share Index over the period 1800 to 2010 are reasonably well described by a Gaussian distribution. Important deviations only emerge at smaller time scales.

distribution is slightly thinner for small returns and seems somewhat fatter for large returns. We return to re-examine this point later.

As a result, assuming Gaussian fluctuations, it also follows that there is no correlation between fluctuations at different times. Hence the normalized autocorrelation function $\langle r(t, \delta t) r(t', \delta t) \rangle$ (eqn (2.12)), is zero everywhere except when $t = t'$. To a good approximation this would have seemed true with the data Bachelier had available. Figure 2.2 illustrates that even today the autocorrelation function for daily log price returns is essentially zero after one day. In Chapters 9 and 10 we shall see that this result seems not to be true for short times–of the order of 30 minutes or so and at these timescales the log-price returns are therefore not independent.

6.2 Evolution of prices

The reader can readily see from eqn (6.7), that the time dependence of log-price changes may be written as

$$\frac{d\langle r(t,\delta t)\rangle}{d\delta t} = v_d. \tag{6.12}$$

Now suppose we would like to obtain, in addition to the time dependence of the log price changes, the time dependence of the price, $s(t)$, itself. Substituting the definition of the log return, $r(t',\delta t) = \ln s(t' + \delta t) - \ln s(t')$, with $t = t' + \delta t$ into the above we obtain

$$\frac{d}{dt}\langle \ln s(t)\rangle = v_d. \tag{6.13}$$

At this stage it is important to note that since we are dealing with random variables it holds that

$$\frac{d\langle s(t)\rangle}{dt} \neq v_d\langle s(t)\rangle. \tag{6.14}$$

We will return to this issue in Chapter 7.

The correct result is obtained by first computing the probability density distribution for the prices and then using this function to compute the first moment of the price. In this way, we obtain $d\langle s(t)\rangle/dt = (v_d + \sigma_0^2/2)\langle s(t)\rangle$ where σ_0^2 is the volatility associated with the fluctuations, as we are now going to show.

To proceed we require that the probabilities for the equivalent fluctuations in prices, $P(s,t)$, and log price returns, $p(r,t)$, are identical. This implies

$$P(s,t)\delta s = p(r,t)\delta r. \tag{6.15}$$

Rewriting eqn (6.11) explicitly in terms of the log price returns we have:

$$p(\ln s(t)) = \frac{1}{\sqrt{2\pi\sigma_0^2 t}}\exp\left(-\frac{(\ln s(t) - v_d t)^2}{2\sigma_0^2 t}\right) \tag{6.16}$$

Using eqn (6.15) we see that the probability density distribution for prices is a *log-normal* distribution,

$$P(s,t) = \frac{\delta r}{\delta s}p(r,t) = \frac{1}{s}p(\ln s) = \frac{1}{s\sqrt{2\pi\sigma_0^2 t}}\exp\left(-\frac{(\ln s - v_d t)^2}{2\sigma_0^2 t}\right), \tag{6.17}$$

with its temporal evolution illustrated in Figure 6.2.

Using this result we can now compute the mean and variance of s.

Exercise 6.2 The reader should show that, after some algebra, the following result is obtained,

$$\langle s(t)\rangle = \int_0^\infty sP(s,t)ds = \exp[v_{d,s}t]. \tag{6.18}$$

Here the *effective drift*, $v_{d,s}$ is not simply equal to v_d but is given by,

$$v_{d,s} = v_d + \sigma_0^2/2. \tag{6.19}$$

Now differentiate with respect to time and we have

$$\frac{d\langle s(t)\rangle}{dt} = v_{d,s}\langle s(t)\rangle = (v_d + \sigma_0^2/2)\langle s(t)\rangle, \tag{6.20}$$

which is indeed consistent with our earlier statement, eqn (6.14).

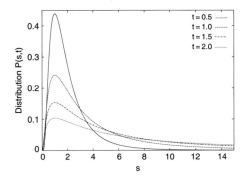

Figure 6.2 The log-normal distribution $P(s,t)$ of eqn (6.17), plotted for several values of t (here we set $\sigma_0^2 = 1, v_d = 1$).

Exercise 6.3 Performing a similar computation as above, the reader should show that the variance of prices is given by

$$\langle s(t)^2 \rangle - \langle s(t) \rangle^2 = \exp[2v_{d,s}t](\exp[\sigma_0^2 t] - 1). \tag{6.21}$$

So we see that whilst average log-prices increase linearly with time at a rate, v_d, prices themselves increase *exponentially* at a rate $v_{d,s} = v_d + \sigma_0^2/2$. This is important from an empirical point of view when drawing conclusions from various data sets plotted as log price and/or price versus time.

The books and papers concerned with theoretical finance and stochastic processes derive these results via rather formal developments in stochastic calculus originally due to Ito. We return to the topic in Chapter 7, where we derive the results of Ito without recourse to stochastic methods. We further shall show that the result of Ito is in fact limited to the special case of Gaussian fluctuations and obtain a more generalized expression valid for more generalized stochastic processes.

6.3 Comparing computer simulations for the geometric Brownian model with real stock data

At this stage it is useful to show some numerical simulations based on the model developed in this chapter and see how well it is suited to describe actual financial data. To this purpose we have written C program code, and analysed both simulation and financial data using the free graphics software Xmgrace[2], which features a number of statistical tools. (Software packages such as Mathematica or Excel are equally suitable.)

The financial data that we have chosen is monthly DJIA data for the period October 1928 to August 2011, i.e. a total of 995 data points, as shown in Figure 6.3(a). From this data we compute 994 log-returns $r(t, 1\text{month}) = \ln(s_{t+1}/s_t)$, which are displayed in Figure 6.3(b).

[2]http://plasma-gate.weizmann.ac.il/Grace/

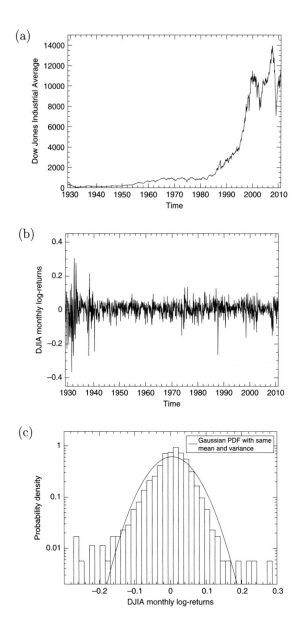

Figure 6.3 The Dow Jones Industrial Average over the period October 1928 to August 2011. (a) Shown is monthly data, i.e. a total of 995 data points, corresponding to the closing price of the first day of every month. (b) The monthly variation of the log-return shows periods of high volatility. (c) The probability distribution of these log-returns is only inadequately described by a Gaussian, which fails to capture the large fluctuations seen in the data.

An empirical probability distribution of these values is obtained by counting the number of log-returns that lie in a certain interval $[r, r+dr]$. The result of this binning procedure is shown in Figure 6.3(c), where we have chosen a logarithmic scale for the ordinate. In such a semi-log plot, a Gaussian distribution, (see eqn (3.10)) turns into an inverted parabola, as can be seen by taking the logarithm, i.e.

$$\ln p(r) = -\frac{\ln(2\pi\sigma^2)}{2} - (r - m)^2/(2\sigma^2). \tag{6.22}$$

The mean m and variance σ^2 of the log-returns can be directly computed from our data in Figure 6.3(b), resulting in $m = 0.0039$ and $\sigma^2 = 0.054^2$. The solid line in Figure 6.3(c) shows a Gaussian distribution, eqn (3.10), with these parameters. Referring to eqn (6.11) this corresponds to $m = v_d\delta t$ and $\sigma^2 = \sigma_0^2\delta t$ where $\delta t = 1$ (month).

This is clearly a poor description for log-returns outside the interval of say [-0.15:0.15]. The probability of having returns of magnitude larger than 0.15 (as feature in Figure 6.3(b)) is drastically underestimated. A closer look at Figure 6.3(b)) also shows that fluctuations are often clustered; we will return to this observation shortly.

Let us now turn to computer simulations of the Brownian motion model. Using a random number generator we create 994 Gaussian distributed random numbers with mean $m = 0.0039$ and variance $\sigma^2 = 0.054^2$, i.e. the values obtained from the DJIA data.

We choose these random numbers to be our log-returns $r(t, 1) = \ln(s_{t+1}/s_t)$ and display them in Figure 6.4(a), where 'time' is simply the sequence of their generation. We can see that very few log-returns are greater than say $3\sigma = 0.16$; large fluctuations relative to the mean value do not occur too often in Gaussian statistics.

The empirical probability distribution of these random numbers is shown in Figure 6.4(b), again with a logarithmic scale for the ordinate. As expected by the construction of our random numbers, they are normally distributed and well described by eqn.(6.22).

Having generated Gaussian distributed log-returns we can finally use these to construct the time series of the price of an imaginary Gaussian asset or an index by simply summing up the computed increments $\ln(s(t-1)/s(t))$ (starting with the log of the October 1928 closing price of the DJIA) and taking the exponential of the value at every time-step. The result is shown in Figure 6.4 for our total of 994 increments, corresponding to 994 unit time steps.

Without detailed analysis, it is easy to imagine that the simulated graphs of Figure 6.4(c) might somehow correspond to the real data displayed in Figure 6.3(a). However, it needs to be remembered that Figure 6.4(c) shows only *one realisation* of a random process. Performing the simulation again with different random numbers, although chosen from the same distribution, will result in a different realisation, such as the one shown in Figure 6.5. This could result in better or less visual similarity to the real DJIA data.

A more important issue concerning the value of our simulation has to do with the character of the fluctuations of the log-returns. As we already mentioned earlier, our random data does not reproduce the clustering observed in the DJIA data. This becomes more obvious when comparing graphs showing the modulus or the square of the log-return as a function of time, as in Figure 6.6. (Note that the clustering is more pronounced in our *daily* DJIA data, shown in Figure 2.1.)

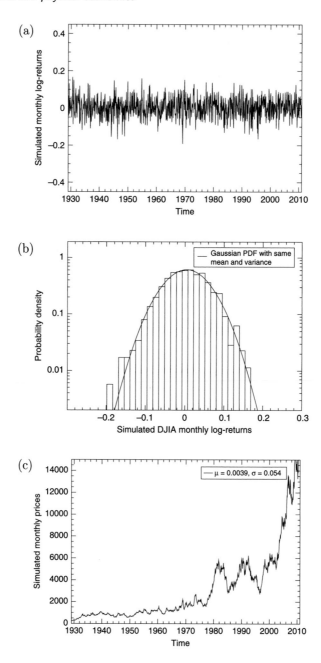

Figure 6.4 Computer simulation using the Brownian motion model of stock prices. (a) Gaussian distributed log-returns with values of mean $m = 0.0039$ and variance $\sigma^2 = 0.054^2$, as obtained from DJIA data from October 1928 to August 2011 (see Figure 6.3). There are hardly any log-returns outside the interval [-0.15:0.15]. (b) Histogram of the simulated log-returns, together with the Gaussian distribution from which they have been chosen. (c) Simulation of a stock index as obtained from consecutively adding up the log-returns of (a) and taking exponentials. A certain visual similarity with the DJIA data of Figure 6.3(a) may not detract from the dissimilarities with respect to the shape of the distribution function of the log-returns and the lack of clustering in this simulation data.

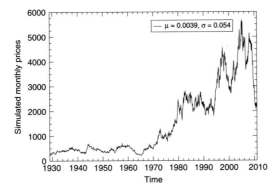

Figure 6.5 Example of a further computer simulation using the same Gaussian distribution as used for the simulation in Figure 6.4 but a different set of random numbers. The visual similarity with the real data shown in Figure 6.4 is less obvious in this case.

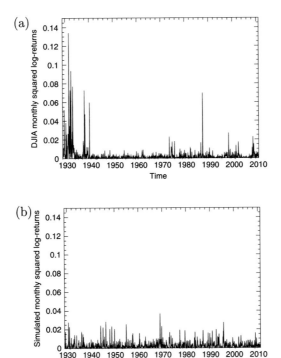

Figure 6.6 The squared values of log-return (a measure of volatility) for both the DJIA monthly data of Figure 6.3(b) and the simulated data of Figure 6.4(a). The volatility of the financial data shows clustering which is absent in the simulated data. This clustering (which is much more pronounced in daily data, see Figure 2.1) is responsible for the slow decay in the autocorrelation function for the squared values of return, $R_r^2(\tau)$ as shown in Figure 6.8(a).

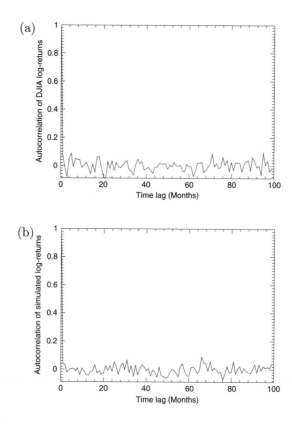

Figure 6.7 Variation of the normalized autocorrelation function $R_r(\tau)$ of log-returns. (a) It is known from minute-by-minute empirical data that $R_r(\tau)$ decays over a time span of approximately 30 minutes, so there are no correlations in the shown *monthly* DJIA data. (b) As expected there are no correlations in the data produced from a computer simulation of the simple Brownian model.

In order to have a closer look at time correlations we will now consider both auto-correlation function of log-returns and the square of log-returns, a measure of volatility. Figure 6.7 shows the normalised autocorrelation function $R_r(\tau)$ (see eqn (2.12)) for the log-price return for both monthly DJIA log-returns and for our simulated log-returns. In both cases R_r fluctuates about zero already after one month (i.e. one time step in the simulation). It is important to note though, that autocorrelations do exist in financial returns. However, they are on time scales of the order of minutes and are thus not visible in the monthly data shown here. We will return to this in Chapter 10 where we discuss minute by minute Dow Jones data.

The situation is different when comparing the normalised autocorrelation function for *squared* returns, $R_{r^2}(\tau)$, as shown in Figure 6.8. In this case the DJIA data shows positive autocorrelation for up to about 40 months, reflecting the clustering in the volatility as we remarked earlier. As expected from the way it has been constructed,

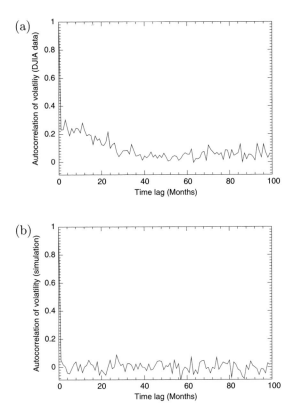

Figure 6.8 Variation of the normalized autocorrelation function $R_{r^2}(\tau)$ of the volatility. (a) The log-returns for monthly DJIA data show positive correlations for up to about 40 months with a sharp drop over the first two months followed by a roughly linear decay. (b) Correlations are absent in the data for log-returns produced using the simple Brownian motion model.

there are no correlations in our simulated data. We had commented on the slowly decaying autocorrelation of the volatility already in Section 2.1.3 and noted that this decay may feature regimes approximated by power laws (see Figure 2.3). The time resolution of our monthly data shown in Figure 6.8(a) is not sufficient to resolve this. We will in Chapter 10 return to this issue when examining minute-by-minute data.

6.4 Issues arising

The comparison of monthly DJIA data with data produced from the simple Brownian motion model of log-returns has shown that while the model serves as a good 'first approximation' of the data, it has severe shortcomings.

- A Gaussian distribution does not describe the observed large values of log-returns that are found in empirical (monthly) data. (*Annual* log-returns are reasonably well described by a Gaussian, see Figure 6.1.)

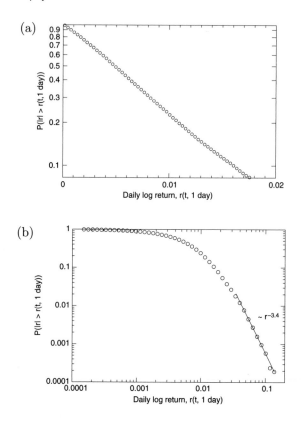

Figure 6.9 (a) Log-linear plot of the cumulative distribution function for *daily* log-price returns for the DJIA. For small returns, the distribution appears linear, suggesting an exponential form would be a good fit. (b) The same data on a log–log plot and suggests a power law might fit the data for large returns. The exponent characterizing the power law decay of such a fit to the 'fat tails' is approximately -3.4.

- The model does not contain the temporal correlations in the volatility which are evident in the empirical data.

Both these observations are much more prominent for daily or minute-by-minute data, as may be illustrated using *daily* DJIA data. Figure 6.9(a) shows that for small returns the cumulative probability distribution $C_>$ of log-returns, and hence also the probability distribution function, decays exponentially (see also McCauley, 2004). However we can also see in Figure 6.9(b) that for large returns the cumulative distribution is suggestive of a power law decay. The exponent characterizing this tail is approximately -3.4 (leading to a probability distribution that decays according to a power law with exponent of about -4.4). This is similar to the tail exponent of about -3 that Liu *et al.* (1999) find for the cumulative distribution function of one minute log-returns for both the S&P500 and individual stocks. It is also in line with our analysis of minute-by-minute DJIA data for the period (1993-2012), presented in Chapter 10.4. This scaling is often referred to as the 'cubic law of returns' (Gabaix *et al.*, 2003).

Today, much detailed data is available and it is apparent to everyone that Gaussian based models do not fit reality. If the distribution functions exhibit power law decay, then large fluctuations can be much more common than would be predicted by a Gaussian distribution. Yet, probably because Gaussians are simple to deal with, many practitioners continue to use models based on the work of Bachelier and Black and Scholes.

Mandelbrot proposed that the empirical data was better described by Lévy distributions. We shall discuss these in Chapter 9 when we examine ideas of scaling of asset data.

Exercise 6.4 Download some stock data for assets from Yahoo (http://uk.finance.yahoo.com/ or http://finance.yahoo.com/) and check the behaviour of the various functions calculated in this chapter.

Exercise 6.5 Example calculation for the expected price of a security
$s(0)$ denotes the price of a certain security on a particular Monday, with $s(n)$ the price of the security at the end of n successive Mondays ($n = 1, 2, 3$, etc).

We model log-price returns $r(n, 1) = \ln(s(n + 1)/s(n))$ as independent, identically distributed random numbers, with a distribution given by eqn (6.16). Writing the mean and standard deviation as $\mu = v_d t$ and $\sigma = (\sigma_0^2 t)^{1/2}$, respectively (time t is discrete in this example), the probability distribution of the log-price returns is given by eqn (6.11).

Using the parameters $\mu = 0.0125$ and $\sigma = 0.055$, estimate the probability that:

(a) The price of the security decreases over the next week.
(b) The price of the security decreases over each of the next two weeks.
(c) The price at the end of two weeks is greater than it is at present.
(d) The price at the end of 20 weeks is less than it is at present.

Answers:

(a) Log returns are normal i.i.d. random variables. Thus the probability that the price will fall in one week (Monday to the following Monday) is:

$$C_<(0) = \int_{-\infty}^{0} p(r, 1)dr = \int_{-\infty}^{0} \frac{dr}{\sqrt{2\pi}\sigma}e^{-(r-\mu)^2/2\sigma^2} = \int_{-\infty}^{-\mu/\sigma} \frac{dy}{\sqrt{2\pi}}e^{-y^2/2}$$

$$= \int_{-\infty}^{0} \frac{dy}{\sqrt{2\pi}}e^{-y^2/2} - \int_{0}^{\mu/\sigma} \frac{dy}{\sqrt{2\pi}}e^{-y^2/2}.$$

Use $\mu/\sigma \sim 0.227 < 1$ to obtain $P_1 = \frac{1}{2} - \frac{0.227}{\sqrt{2\pi}} \sim 0.46$.

(b) The probability of a decrease over each of the next two weeks is given by $C_<(0)^2 \sim 0.21$.

(c) Probability at the end of two weeks that the price has changed by $z = x_1 + x_2$ is $p_2 = \int_{-\infty}^{\infty} \delta(z - [x_1 + x_2])p_G(x_1)p_G(x_2)dx_1dx_2$. This is a convolution of two Gaussian functions, which is also a Gaussian where $\sigma \to 2\sigma$. Thus $p_2 = \frac{1}{\sqrt{2\pi 2\sigma}}e^{-(r-\mu)^2/2(2\sigma)^2}$. Probability that after two weeks price has risen is

$$P_>(0,2) = \int_0^{\infty} p_2(y)dy = \int_0^{\infty} \frac{dy}{\sqrt{2\pi 2\sigma}}e^{-(y-\mu)^2/2(2\sigma)^2}$$

$$= \int_{-\mu/2\sigma}^{\infty} \frac{du}{\sqrt{2\pi}}e^{-y^2/2} = \int_{-\mu/2\sigma}^{0} \frac{du}{\sqrt{2\pi}}e^{-y^2/2} + \int_0^{\infty} \frac{du}{\sqrt{2\pi}}e^{-y^2/2}$$

$$\sim 0.045 + 0.5 = 0.55.$$

(d) Probability that at the end of 20 weeks, the price has risen is

$$P_>(0,20) = \int_0^{\infty} p_{20}(y)dy = \int_0^{\infty} \frac{dy}{\sqrt{2\pi 20\sigma}}e^{-(y-\mu)^2/2(20\sigma)^2}$$

$$= \int_{-\mu/20\sigma}^{\infty} \frac{du}{\sqrt{2\pi}}e^{-y^2/2} = \int_{-\mu/20\sigma}^{0} \frac{du}{\sqrt{2\pi}}e^{-y^2/2} + \int_0^{\infty} \frac{du}{\sqrt{2\pi}}e^{-y^2/2}$$

$$\sim 0.03 + 0.5 = 0.503.$$

7
Generalized diffusion processes and the Fokker-Planck equation

Besides, I had a contempt for the uses of modern natural philosophy. It was very different when the masters of the science sought immortality and power; such views, although futile, were grand: but now the scene was changed. The ambition of the inquirer seemed to limit itself to the annihilation of those visions on which my interest in science was chiefly founded. I was required to exchange chimeras of boundless grandeur for realities of little worth.

Mary W. Shelley, Frankenstein.

The Chapman-Kolmogorov equation, eqn (4.15), derived in Chapter 4, is an integral equation. In this chapter, where we limit our discussion to Markov processes (see Section 4.2), we show how it can be reduced to the Fokker-Planck partial differential equation, using the concept of n–th order diffusion constants. These diffusion constants find applications for example when computing the evolution of the average of a random variable (Ito's calculus lemma), and we will illustrate this to obtain the average price of a Brownian asset as a function of time. Finally, we will introduce a more general Langevin equation than the one introduced in Chapter 5 in order to study functions of random variables.

Throughout this chapter, whose content forms the basis of Chapters 8 on options and 10 on more realistic models of asset fluctuations, our approach essentially follows that by Lax *et al.* (2006).

7.1 Introduction of n-th order diffusion constants

First we need to introduce the concept of generalized diffusion processes and obtain Ito's calculus lemma. This lemma can be a source of much confusion and not all financial text books are clear in their exposition. The approach here, as developed by Lax *et al.* (2006), has the merit of being more general and, in the opinion of these authors, conceptually simpler.

Setting $t' = t + \Delta t$, and thus $x' = x(t') = x(t + \Delta t)$, and $x_0 = x(t = t_0)$, we may write the Chapman-Kolmogorov equation as

$$p(x', t + \Delta t | x_0, t_0) = \int p(x', t + \Delta t | x, t) p(x, t | x_0, t_0) dx. \qquad (7.1)$$

In order to proceed we assume that moments of the transition probability $p(x', t + \Delta t | x, t)$, exist and can be expressed as a Taylor series in Δ. Thus for $n \geq 1$ and infinitesimal Δ we obtain for the n–th moment

$$\int (x' - x)^n p(x', t + \Delta t | x, t) dx' = n! D_n(x, t) \Delta t + O(\Delta t^2), \tag{7.2}$$

where D_n is called the *n-th order diffusion constant*, defined as,

$$D_n(x, t) = \frac{1}{n!} \lim_{\Delta t \to 0} \int \frac{(x' - x)^n p(x', t + \Delta t | x, t) dx'}{\Delta t}$$

$$= \frac{1}{n!} \lim_{\Delta t \to 0} \frac{\langle [x(t + \Delta t) - x(t)]^n \rangle_{x(t)=x}}{\Delta t}. \tag{7.3}$$

From a physical perspective we are assuming here that the fluctuations in x can be extremely rapid over the small time interval Δt. Therefore if for any specific value of n, the moment proves to be $\sim \Delta t^k$ where $k > 1$, then the associated diffusion coefficient is zero.

This definition is central to the development of a theory of random processes. Random variables differ from non-random variables in an essential way. For a non-random variable, $(x' - x)^n$ is a smooth function and is proportional to Δt^n. Hence as $\Delta t \to 0$, $(x' - x)^n / \Delta t \to 0$ for $n > 1$. But for a fluctuating random variable this does not hold, since there may be n-th moments that are proportional to Δt. This leads to non-zero n-th order diffusion constants.

The first order diffusion coefficient, D_1,

$$D_1(x, t) = \lim_{\Delta t \to 0} \left\langle \frac{\Delta x}{\Delta t} \right\rangle_{x(t)=x} \tag{7.4}$$

is usually called the *drift*, since it seems reasonable to expect the average value of x to satisfy an equation of the following form,

$$\frac{d \langle x(t) \rangle_{x(t_0)=x_0}}{dt} = \langle D_1(x, t) \rangle = \int D_1(x, t) p(x, t | x_0, t_0) dx. \tag{7.5}$$

The second coefficient, D_2, is given by

$$D_2(x, t) = \frac{1}{2} \lim_{\Delta t \to 0} \left\langle \frac{(\Delta x)^2}{\Delta t} \right\rangle_{x(t)=x} \tag{7.6}$$

and is related to the familiar diffusion coefficient that arises in Brownian motion. From eqns (5.22) and (5.23) we obtain

$$D_2 \simeq \frac{\alpha}{2\lambda^2} = k_B T / \zeta = D. \tag{7.7}$$

Diffusion coefficients of order greater than $n = 2$ have not usually been used in physics.

7.2 Evolution of the average of a random variable

If in eqn (7.1) we integrate over the variable x_0 then we obtain an equation for the probability distribution, $p(x, t)$:

$$p(x', t + \Delta) = \int p(x', t + \Delta | x, t) p(x, t) dx. \tag{7.8}$$

Now consider a general function $M = M(x, t)$ of our random variable, x. The average motion is obtained by integrating $M(x', t + \Delta t)$ against $p(x', t + \Delta)$,

$$\langle M(x', t + \Delta t) \rangle = \int M(x', t + \Delta t) p(x', t + \Delta t) dx'. \tag{7.9}$$

A Taylor expansion of the function M results in

$$M(x', t + \Delta t) = M(x, t) + \frac{\partial M}{\partial t} \Delta t + \frac{1}{n!} \sum_{n=1}^{\infty} (x' - x)^n \frac{\partial^n M}{\partial x^n}, \tag{7.10}$$

where we have curtailed the expansion at first order in t, but consistent with our assumption above that diffusion coefficients D_1, D_2 and higher orders D_n can exist, retained all terms in the expansion of $x - x'$.

The LHS of eqn (7.9) is evaluated by inserting both this Taylor expansion and the expression for $p(x', t + \Delta t)$. Using the normalization $\int p(x', t + \Delta t | x, t) dx' = 1$ and the definition of the n-th order diffusion coefficients, eqn (7.3), then results in

$$\frac{d \langle M(x, t) \rangle}{dt} = \left\langle \frac{\partial M}{\partial t} \right\rangle + \sum_{n=1}^{\infty} \left\langle D_n(x, t) \frac{\partial^n M(x)}{\partial x^n} \right\rangle, \tag{7.11}$$

in the limit $\Delta t \to 0$.

Exercise 7.1 The reader may wish to carry out this derivation as described above.

The conditional expectation of $M(x)$, given when $x(t) = x$, is obtained by setting $p(x, t) = \delta(x(t) - x)$. This yields

$$\frac{d \langle M(x) \rangle}{dt} \bigg|_{M(t)=M} = \frac{\partial M}{\partial t} + \sum_{n=1}^{\infty} D_n(x, t) \frac{\partial^n M(x)}{\partial x^n} \bigg|_{x(t)=x}. \tag{7.12}$$

If we retain in the summation only the first and second terms we obtain

$$\frac{d \langle M(x) \rangle}{dt} \bigg|_{M(t)=M} = \frac{\partial M}{\partial t} + D_1(x, t) \frac{\partial M(x)}{\partial x} \bigg|_{x(t)=x} + D_2(x, t) \frac{\partial^2 M(x)}{\partial x^2} \bigg|_{x(t)=x}. \tag{7.13}$$

Equation (7.13), which is valid only for Gaussian random variables (since we assume that the second moment is finite), is referred to as *Ito's calculus lemma* and has played an important role in the development of theories of stock prices, as well as in other areas of physics where Gaussian stochastic processes are important.

7.3 Application to simple stock price model

As an application of this lemma, let us return to the geometric Brownian motion model of stock prices introduced in Section 6.2. The average of the returns $r(t, \delta t)$ was given by $d\langle r(t, \delta t)\rangle/d\delta t = v_d$ where v_d is a drift velocity and the aim was to compute $d\langle s\rangle/dt$ for the price $s(t' + \delta t) = s(t')\exp(r(t', \delta t))$.

Now let $s(r)$ take the role of the function $M = M(x)$ in the discussion above. The variation of the average price may then be obtained using Ito's lemma, as $\frac{d\langle s(r)\rangle}{dt}|_{s(t)=s} = D_1\frac{ds}{\partial r} + D_2\frac{\partial^2 s}{\partial r^2}$. Since in this example the partial derivatives with respect to r simply reproduce s, we obtain $\frac{d\langle s\rangle}{dt}|_{s(t)=s} = (D_1 + D_2)s$. From the definition of the diffusion constants, eqn (7.3), we can identify D_1 and D_2 as $D_1 = v_d$ and $D_2 = \sigma_0^2/2$ (see eqns (6.7) and (6.9), respectively) to obtain

$$\frac{d}{dt}\langle s\rangle|_{s(t)=s} = v_{d,s}s \text{ where } v_{d,s} = v_d + \sigma_0^2/2, \tag{7.14}$$

consistent with our earlier result, eqn (6.20), which was obtained for the non-conditional average. While the average return increases linearly with time, the average price increases exponentially. The difference in the two growth rates is due to the stochastic nature of the process studied, i.e. the existence of non-zero diffusion coefficients. This point is obviously important when computing numerically values using empirical data for log-price and return values.

7.4 The Fokker-Planck equation

In the case where M does not depend explicitly on time, eqn (7.11) can be written as follows,

$$\int\left[\sum_{n=1}^{\infty} D_n(x, t)p(x, t)\frac{\partial^n M(x)}{\partial x^n} - M(x)\frac{\partial p(x, t)}{\partial t}\right]dx = 0. \tag{7.15}$$

Integrating by parts (n times for the n-th term in the summation) and noting that the probability density P vanishes as x→ ∞, we obtain

$$\int\left[\sum_{n=1}^{\infty}(-1)^n\left(\frac{\partial^n}{\partial x^n}\right)[D_n(x, t)p(x, t)] - \frac{\partial p(x, t)}{\partial t}\right]M(x)dx = 0. \tag{7.16}$$

Since $M(x)$ is arbitrary, the term within the square brackets must be zero. In this way we obtain the *generalized Fokker-Planck equation*:

$$\frac{\partial p(x, t)}{\partial t} = \sum_{n=1}^{\infty}(-1)^n\left(\frac{\partial^n}{\partial x^n}\right)[D_n(x, t)p(x, t)]. \tag{7.17}$$

The Fokker-Planck equation often found in the physics literature is obtained by truncating the series at $n = 2$.

7.5   Application: The Maxwell-Boltzmann distribution of velocities

Now recall the discussion of Brownian motion in Chapter 5 where we obtained expressions for the average velocity, eqn (5.7) and the mean square fluctuations, eqn (5.10),

$$\langle \Delta v(t) \rangle_{v_0} = \langle v(t) - v_0 \rangle = v_0(e^{-\lambda \Delta t} - 1),$$

$$\langle \Delta v^2(t) \rangle = \langle v^2(t) - v_0^2 e^{-2\lambda \Delta t} \rangle_{v_0} = \frac{\alpha}{2\lambda}(1 - e^{-2\lambda \Delta t}). \tag{7.18}$$

If we expand the RHS of both these equations, and retain only terms of first order in Δt, we obtain $\langle \Delta v(t) \rangle_{v_0} \approx -v_0 \lambda \Delta t$ and $\langle \Delta v^2(t) \rangle \approx \alpha \Delta t$. Recognizing that within this approximation $v_0 \sim v$, and using eqns (7.4) and (7.6), we obtain

$$D_1(v) = -\lambda v,$$

$$D_2(v) = \alpha/2. \tag{7.19}$$

Substituting these expressions into the Fokker-Planck eqn (7.17) and truncating the series at $n = 2$, we obtain

$$\frac{\partial p(v,t)}{\partial t} = \left(\frac{\partial^2}{\partial v^2} \right) [D_2(v,t)p(v,t)] - \frac{\partial}{\partial v}[D_1(v,t)p(v,t)] =$$

$$= \frac{\alpha}{2} \frac{\partial^2 p(v,t)}{\partial v^2} + \lambda \frac{\partial [vp(v,t)]}{\partial v} \tag{7.20}$$

The stationary solution, $p^*(v)$ is obtained when the LHS of this equation is set to zero. Since the probability distribution also tends to zero in the limit $v \to \pm\infty$ we readily obtain

$$p^*(v) = \frac{1}{N} \exp(-\frac{\lambda}{\alpha}v^2), \tag{7.21}$$

where N is a normalization factor. Identifying λ/α with $m/(2k_bT)$, this as we expect, is the Maxwell distribution of velocities for one dimensional motion, which we already derived in Section 5.2 (see eqn (5.14)), when discussing Brownian motion.

Exercise 7.2 The reader may wish to derive the above expression for $p^*(v)$.

7.6   The Langevin equation for a general random variable

In order to be able to deal with options, as will be discussed in Chapter 8, and to relate to the discussion in Chapter 5, we will now consider more general Langevin equations of the form

$$\frac{dx}{dt} = F(x,t) + G(x(t))f(t), \tag{7.22}$$

where F and G are general functions of the random variable x. Again our approach will be guided by the work of Lax *et al.* (2006).

First note that the random function $f(t)$ has the following properties,

$$\langle f(t) \rangle = 0, \text{ and } \langle f(t)f(t') \rangle = 2\delta(t - t'), \tag{7.23}$$

and higher order averages, e.g. $\langle f(t)f(t')f(t'') \rangle$ are assumed to be zero. This essentially limits what follows to Gaussian processes. Note also that eqn (7.22) does not depend on derivatives higher than first order. This implies that the stochastic process is Markovian and we can introduce diffusion coefficients, as defined at the beginning of this chapter, see eqn (7.3).

Now we ask what is the corresponding Langevin equation for a function $M(x,t)$ of the random variable $x(t)$?

To proceed we rewrite eqn (7.22) as an integral equation,

$$\Delta x(t) = \int_t^{t+\Delta} F(x(t))dt + \int_t^{t+\Delta} G(x(t))f(t)dt, \tag{7.24}$$

which we can now solve *by iteration*. Retaining only terms of order Δt or f^2, but not $f\Delta t$ or higher order gives

$$\Delta x = F(x)\Delta t + G(x)\int_t^{t+\Delta t} f(t')dt' + G(x)\frac{\partial G(x)}{\partial x}\int_t^{t+\Delta t} f(t')dt' \int_t^{t'} f(t'')dt''. \tag{7.25}$$

Exercise 7.3 The reader might wish to derive the above result. $G(x(t))$ and $\frac{\partial G(x(t))}{\partial x}$ are slowly varying functions, computed at time t, and can thus be placed outside the integral.

Now we can form the moments, D_n. For $n > 2$, using the properties of the random force, $f(t)$, we obtain $D_n = 0$. For $n = 2$, we obtain

$$D_2(x,t) = \lim_{\Delta t\to 0}\frac{\langle(\Delta x)^2\rangle}{2\Delta t} = \lim_{\Delta t\to 0}\frac{G^2(x)}{2\Delta t}\int_t^{t+\Delta t} dt' \int_t^{t+\Delta t} dt'' \langle f(t')f(t'')\rangle = G^2(x). \tag{7.26}$$

For $n = 1$ we have

$$D_1(x,t) = \lim_{\Delta t\to 0}\frac{\langle\Delta x\rangle}{\Delta t} = F(x) + \frac{G(x)}{\Delta t}\frac{\partial G(x)}{\partial x}\int_t^{t+\Delta t} dt' \int_t^{t'} dt'' \langle f(t')f(t'')\rangle$$

$$= F(x) + \frac{G(x)}{\Delta t}\frac{\partial G(x)}{\partial x}\frac{1}{2}\int_t^{t+\Delta t} dt' \int_t^{t+\Delta t} dt'' \langle f(t')f(t'')\rangle. \tag{7.27}$$

The second line follows by first assuming that $\langle f(t)f(t')\rangle$ may be treated as a continuous symmetric function. We may then change the limits of integration in the second integral from t, t' to $t, t+\Delta t$ and divide this second integral by $1/2$, so that we still only integrate over a triangle in the $t' - t''$ plane and not a square.

We can now use eqn (7.23) to obtain

$$D_1(x,t) = F(x) + G(x)\frac{\partial G}{\partial x}. \tag{7.28}$$

With these relationships between D_1 and D_2 and F and G, we can move from the Fokker-Planck equation over to the equivalent Langevin representation and vice versa.

In books on financial mathematics one frequently comes across statements to the effect that 'these random variables, x and M, etc. do not transform according to the usual laws of calculus', which can be very confusing. We now show that the method developed here leads to random variables that *do* transform according to the usual laws of calculus.

To see this, we first need to consider a general function $M = M(x(t), t)$ of a random variable $x(t)$ that itself obeys the Langevin eqn (7.22) and determine how M evolves in time.

From Ito's calculus lemma, eqn (7.13), we first note that the drift $A(M, t) = \langle dM/dt \rangle|_{M(t)} \equiv D_1(M, t)$ associated with M is

$$A(M, t) = \langle dM/dt \rangle|_{M(t)} = \frac{\partial M}{\partial t} + D_1 \frac{\partial M}{\partial x} + D_2 \frac{\partial^2 M}{\partial x^2}. \tag{7.29}$$

The Langevin equation we require for M has the following form

$$\frac{dM}{dt} = F(M) + G(M)f(t), \tag{7.30}$$

with drift $F(M)$ and stochastic coefficient $G(M)$.

The task is then to express $F(M)$ and $G(M)$ in terms of $F(x)$ and $G(x)$, respectively, i.e. the drift and stochastic coefficient in the Langevin equation for x, eqn (7.22).
Writing $A(M) = F(M) + G(M)\frac{\partial G}{\partial x}$ (compare with eqn (7.28)), we obtain

$$F(M) = A(M) - G(M)\frac{\partial G}{\partial M} = \frac{\partial M}{\partial t} + D_1 \frac{\partial M}{\partial x} + D_2 \frac{\partial^2 M}{\partial x^2} - G(M)\frac{\partial G}{\partial M}, \tag{7.31}$$

where we have used eqn (7.29) for the drift A.

In terms of M, G is defined as follows,

$$G(M) = \left[\frac{\langle [M(x + \Delta x) - M(x)]^2 \rangle}{2\Delta t} \right]^{1/2} = \frac{\partial M}{\partial x} \left[\frac{\langle \Delta x^2 \rangle}{2\Delta t} \right]^{1/2} = \frac{\partial M}{\partial x} G(x), \tag{7.32}$$

where we have Taylor-expanded the term in square brackets.

Using this result for $G(M)$, the last term on the RHS of eqn (7.31) may now be re-expressed as follows,

$$G(M)\frac{\partial G(M)}{\partial M} = \frac{\partial M}{\partial x} G(x)\frac{\partial}{\partial M} \left[\frac{\partial M}{\partial x} G(x) \right] = G^2(x)\frac{\partial^2 M}{\partial x^2} + G(x)\frac{\partial M}{\partial x}\frac{\partial G(x)}{\partial x}. \tag{7.33}$$

Exercise 7.4 The reader may like, as an exercise, to check this expression.

Substituting the expressions for $G(M)\frac{\partial G(M)}{\partial M}$, D_1(eqn (7.28)) and D_2 (eqn (7.26)) into eqn (7.31), we finally obtain after a little manipulation

$$F(M) = \frac{\partial M}{\partial t} + \frac{\partial M}{\partial x} F(x). \tag{7.34}$$

The Langevin equation for M, eqn (7.30), may thus be rewritten as

$$\frac{dM}{dt} = \frac{\partial M}{\partial t} + \frac{\partial M}{\partial x} F(x) + G(x)\frac{\partial M}{\partial x} f(t) = \frac{\partial M}{\partial t} + \frac{\partial M}{\partial x}\frac{dx}{dt}, \tag{7.35}$$

where we have used the Langevin equation for x, eqn (7.22).

We now see that this transform, from one Langevin equation for M to another for x follows using the ordinary laws of calculus. No additional terms or higher derivatives are present.

However, note that the average $\langle dM/dt \rangle$ has contributions not only from the first term $\langle F(M) \rangle$, but also the second term, $\langle G(M)f(t) \rangle$, since (from eqns (7.31) and (7.35) the following holds,

$$\langle G(M)f(t) \rangle = \langle G(M) \frac{\partial G(M)}{\partial M} \rangle. \tag{7.36}$$

This is zero only if G is a constant, independent of M. For the conditional average with $M(t) = M$ we have

$$\langle G(M)f(t) \rangle_{M(t)=M} = G(M) \frac{\partial G(M)}{\partial M}. \tag{7.37}$$

7.7 Application to geometric Brownian motion model of stock prices

Let us return to the model of Chapter 6 to study the evolution of the stock price $s(t)$. The log-price returns r are determined by the Langevin equation

$$\frac{dr}{dt} = v_d + \sigma f(t), \tag{7.38}$$

where v_d is the mean return (drift velocity) and σ is the volatility associated with the random process $f(t)$. Both v_d and σ are now assumed to be constant and independent of the random variable r. Note that since the average over the fluctuations is zero (see eqn (7.23)), we obtain $\frac{d\langle r \rangle}{dt} = v_d$, as in eqn (6.12).

Since the Langevin equation obeys the ordinary laws of calculus, we can write the Langevin equation (eqn (7.35)) for the price, by setting $M = s(r) = s_0 \exp r$. Unlike the results found in many books on financial mathematics, we simply have $dr = ds/s$, according to the rules we learn in school and obtain

$$\frac{ds}{dt} = v_d s + \sigma s f(t). \tag{7.39}$$

In order to work out the average $\langle ds/dt \rangle$, we need to determine $\langle \sigma s f(t) \rangle$. From eqn (7.32) we obtain for the fluctuation term for s, $\sigma(s) = \sigma(r)\partial s/\partial r = \sigma s$. The conditional average is then obtained from eqn (7.37) as $\langle \sigma s f(t) \rangle|_{s(t)=s} = \sigma(s)\partial\sigma(s)/\partial s = \sigma^2 s$, resulting in

$$\frac{d\langle s \rangle}{dt}|_{s(\delta t)=s} = (v_d + \sigma^2)s. \tag{7.40}$$

Setting $\sigma^2 = \sigma_0^2/2$ we note that we had obtained this result earlier, eqn (7.14), using Ito's calculus lemma.

The many books published on financial mathematics usually evaluate stochastic integrals slightly differently to the approach used here. Their approach due originally to Ito assumes the function $G(M)$ is evaluated at the beginning of the time of integration.

This leads to the average $\langle Gf \rangle = 0$, differing from eqn (7.36), and a resulting Langevin equation that does not follow the usual laws of calculus. Assuming, as we do, that averages may be treated as limits of continuous analytic functions is a more intuitive approach for a physicist (In technical terms, we use a *Stratonovich* representation of a stochastic integral, which evaluates $G(M)$ as an average at beginning and endpoints of each integration interval). The effect of Ito that arises from the strong fluctuations is then routed into the third term on the RHS of eqn (7.13) and second term on the RHS of eqn (7.40) in the example above.

In mathematics and mathematical finance literature, it is common to describe stochastic processes by

$$dx = a(x, t)dt + b(x, t)dW, \tag{7.41}$$

where $a(x, t)$ and $b(x, t)$ are general functions of the random variable x and time t, and dW describes a so-called Wiener process (cf. the equivalent Langevin equation, eqn (7.22)). The change of the random variable dW in an infinitesimal time interval, dt, is given by

$$dW(t) = \int_t^{t+dt} dt' f(t'), \tag{7.42}$$

where $f(t)$ is the random function defined by eqn (7.23).

In this formulation, the geometric Brownian motion model for stock prices is written as

$$ds = v_d s dt + \sigma s dW, \tag{7.43}$$

resulting in the following stochastic differential equation for the log-return (Voit, 2001)

$$dr = (v_d - \frac{\sigma^2}{2})dt + \sigma dW. \tag{7.44}$$

In the next chapter we shall develop the theory of option prices for which the results developed in this chapter are essential.

8
Derivatives and options

Listen to this, you who grind the destitute and plunder the humble, you who say, 'When will the new moon be over so that we may sell corn? When will the Sabbath be past so that we may open our wheat again, giving short measure in the bushel and taking overweight in the silver, tilting the scales fraudulently, and selling the dust of the wheat; that we may buy the poor for silver and the destitute for a pair of shoes?'
   Book of Amos 8: 4-6

The use of derivatives can be traced back thousands of years. Mesopotamian farmers routinely sold the following year's grain harvest due for delivery at the time of harvesting at a 'future' price, agreed today. The actual future price went up and down according to the market. For example, a drought during the summer would give rise to fears of a poor harvest and the price would rise. Equally a good summer creating the opinion that the harvest would be good would lead to the price remaining at reasonable levels. The profit or loss on the sale of the harvest was, however, determined by the future price agreed the previous year.

Similar transactions are now widely used by business. For a fee, an agreement is made to engage, at some point in the future, in a trade at a fixed price. The fee is the price of the so-called derivative contract which clearly should depend in some way on the price of the underlying asset. In 1992, one US bank had option contracts totalling over $1000 billion. The total value of contracts in play at that time was estimated at $10,000 billion. Today the figure is likely to be much, much greater.

Gold mine owners and jewellers, wheat farmers and bakers, oil companies and airlines all seek to reduce uncertainty in the financial state of their businesses by trying to fix the price of the underlying commodity. As a result, they are all natural buyers and sellers of derivative contracts. Nowadays, others also take advantage of the derivatives market to not only hedge or insure against adverse future price changes, but also to speculate and gamble on future price changes—an inevitable by-product of these underlying real world processes.

8.1 Forward contracts and call and put options

The contract agreed by the Mesopotamian farmer is an example of a *forward contract*, where person A agrees to buy and B agrees to sell an asset at a specific price, K (the forward price) on a specific date, T (the delivery date). The financial community then says that A has taken a *long* position and B has taken a *short* position. Since the price will fluctuate, the actual price, $s(T)$, at the delivery date, T, may be considerably different to the agreed forward price, K, and the payoff or profit may be positive or negative as shown in Figure 8.1.

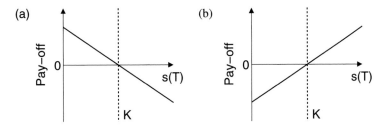

Figure 8.1 Payoff for a forward contract. (a) shows the payoff as a function of the stock price $S(T)$ at the delivery date due to the short position of the seller; (b) shows the corresponding payoff due to the long position of the buyer. K is the agreed forward price.

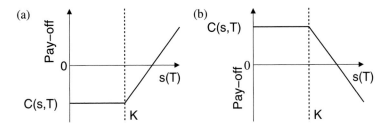

Figure 8.2 Payoff for the buyer of a call option (a) and for a put option (b). $C(S,T)$ is the price paid for the option to put/sell it at time T for the specified price, K.

Note that whatever profit is made by one party, is lost by the other. There is, in this sense, symmetry in the nature of the payoff.

There are many other examples of derivatives that go under names such as futures, swaps, straddles and options. We shall restrict ourselves here to *options*.

A European *call option* gives the holder the right to buy an asset at a fixed price, K at a specific time T in the future. The option is secured for the option price or fee, $C(s,T)$, see Figure 8.2(a). There is a complementary *put option* that for a fee gives the holder the right to sell an asset for a fixed price, K, at a specific future time, T, as illustrated in Figure 8.2(b).

The European option can only be exercised at the exercise or expiry time T; however, an American call option can be exercised at any time up to the expiry time T. There are other more exotic options that offer different ways for exercising the contract.

Unlike the futures contract, now there is no symmetry between the payoff for the holder of the contract, the buyer, and the seller of the contract. The buyer of a call option pays the option price or fee C at time $t = 0$ and acquires the right to buy at $t = T$. The seller receives cash but faces potential liabilities when the option is exercised at time T.

The following questions immediately arise. What is a fair price for an option and how can a seller of a call option minimize the risk associated with the potential liability?

8.2 A simple example illustrating the effect of options

To get more insight into the effect of options let us consider a simple and hypothetical example. Suppose in three months' time, using a call option, we may purchase one share in Acme Ltd for 2.50. Now suppose this particular share can only take on a few discrete values, namely 2.29, 2.50, or 2.71. So we have three possible scenarios:

Scenario 1. In 3 months Acme Ltd trades at 2.71. We can then exercise the option: buy for 2.50; sell at 2.71; profit is 0.21.

Scenario 2. In 3 months Acme Ltd trades at 2.29. In this case we simply let the option lapse. There is no point in buying a stock at 2.50 and then selling it in the market at 2.29 thus losing 0.21 in the process.

Scenario 3. In three months' time Acme Ltd continues to trade at 2.50. Again, we would let the option lapse, since using our option we could only buy and sell at the same price, thus making no profit on the deal. (Indeed, we would lose as a result of both buying and selling administration costs).

Assuming these three scenarios are all equally likely, then the expected profit is $0.21 \times 1/3 + 0 \times 1/3 + 0 \times 1/3 = 0.07$. So it is reasonable to assume the cost of such an option should be 0.07. If Scenario 1 comes to pass then we gain 0.14—a profit of 200%! For each of the other two scenarios we lose the initial cost of the option, 0.07—a loss of 100% of our initial investment.

If we had bought the *stock* at 2.50 (instead of the option), then after three months we would, given Scenario 1, make a profit of $0.21/2.50 \times 100$—a profit of 8.4%. Clearly options allow substantial gearing of profits to be made for smaller outlays. Equally the losses can also be greater!

However, one can never know the future price of an asset. Since it is determined by a random variable, at best, we can only hope to compute the probability that the price will take a particular value. So how then do we compute option prices? The first serious attempt to answer this question was provided by the theory of mathematician Fisher Black and economist Myron Scholes in 1973. This work was further expanded upon by the economist Robert C. Merton (1973). Scholes and Black were awarded the Nobel prize for the work in 1997, but unfortunately Fisher Black had died by this time. Publication of the Black and Scholes theory coincided with the introduction of the first programmable pocket calculators and traders soon found that they could programme the various formulae into these machines.

Before presenting in Section 8.3 the theory which assumes that the log-price returns of the underlying asset are Gaussian distributed (cf. Chapter 6), let us have a brief look at some empirical data for option prices.

Today, option prices for a range of stocks are regularly published in the financial press and real time prices are quoted on numerous websites that trade option contracts. Examples are shown in Table 8.1.

Option prices for both calls (C) and puts (P) are quoted not just for stocks, but also for the FTSE index itself. In this case, the values can be interpreted as a kind of

Table 8.1 One, three and six month option prices for four different UK companies and two different strike prices published on 22 November 2003. The columns can be understood as follows. The price of the stock on 22 November 2003 is the number below the name of each stock. So for example Br Air on 22 November 2003 was 214. Assuming the price will fall, the cost of a call option to buy Br Air at a price of 200 in January is 21.25; alternatively one might think the price would rise and the cost of a call option to buy Br Air at 220 in January is 11.75.

Option		$\cdots$	Calls	$\cdots$	$\cdots$	Puts	$\cdots$
		Jan	Apr	Jun	Jan	Apr	Jun
Br Air	200	21.25	32.25	32.26	8.25	14.5	20.25
214	220	11.75	22	25.75	16.75	24	30.25
Centrica	180	14.4	20	22	3.5	7.5	10.5
190	200	4.5	10	11	13.5	17	19.5
Option		Dec	Mar	Jun	Dec	Mar	Jun
ARM	100	12	18	22	2.5	8	11
109	110	6	12	17	6.5	12	16
Unilever	500	14.5	30	38.5	4	15.5	26.5
509	550	0.5	8	15	41	44.5	55

bet against the price rising or falling. Some typical values are shown in Table 8.2 and graphed in Figure 8.3.

From such data we can deduce a boundary condition that applies to option prices. If, at the maturity time T, the stock price s has risen above the strike price K, the call option must be priced at $C(s,T) = s(T) - K$, so that whilst a caller could buy the option at time T for price C, the profit $s - K$ is equal to the cost or value of the call option. However, if the stock price s has fallen below the strike price K, then the call option is not exercised, since it would result in a loss and the value of the option is worthless, or $C(s,T) = 0$ if $s \leq K$. The boundary condition is continuous at $s(T) = K$. This is frequently summarized as

$$C = \max\{s(T) - K, 0\}. \tag{8.1}$$

8.3   The theory of Black and Scholes

Using our approach developed in Chapter 7, the derivation of the theory of Black and Scholes is straightforward, as will now be presented, together with a discussion of the solution of the Black-Scholes equation.

8.3.1   The rational portfolio

The first aspect of the problem is to define the asset portfolio that exists during the lifetime of the derivatives contract. If I sell a contract that allows someone to buy from me at a time T in the future then I shall receive immediately at present time, t, an

Table 8.2 Option prices for the FTSE index quoted on 23 November 2003. One can see that these options are essentially bets on a belief that the FTSE index will in the months following November, either rise, in which case one buys a call option, or fall, in which case one buys a put option. It is necessary to pay £10 per index point change from the 21 November value. (FTSE (21 November 2003) = 4319.)

	4025		4125		4225		4325	
	Call	Put	Call	Put	Call	Put	Call	Put
Nov	287	0.25	187	0.25	86.5	0.25	0.25	13.5
Dec	310	17	220	27	139	45	73.5	80
Jan	333	30.5	250	46.5	174	70	109	105
Feb	362	50.5	282	69	209	95	145	131
Mar	361	76	286	100	219	132	159	171
	4425		4525		4625		4725	
	Call	Put	Call	Put	Call	Put	Call	Put
Nov	0.25	114	0.25	214	0.25	314	0.25	414
Dec	31	137	11.5	218	4	310	1	407
Jan	60.5	155	31.5	226	15.6	310	6.5	400
Feb	94	175	56.5	241	31.5	314	17.5	400
Mar	111	221	71.5	281	45	353	26.5	433

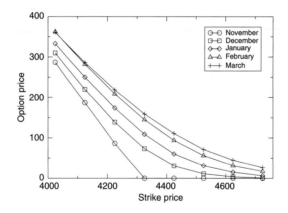

Figure 8.3 Option price as a function of exercise price for the FTSE call option in November 2003 for different maturity dates. The characteristics at the expiry date are illustrated by the November data with a strike price of 4319 at 21 November 2003, the third Friday in the month. (The data is taken from Table 8.2.)

amount, C, the value of the option. However, to cover my potential liabilities I should also immediately purchase a number, n, of shares of the asset in the market place, at a total cost of $ns(t)$. The value ϕ of my portfolio is then

$$\phi = -C(s, T) + ns(t). \tag{8.2}$$

How do I choose the number n? Consider the variation in the portfolio value over a short time, Δt,

$$\Delta \phi = \phi(t + \Delta t) - \phi(t) = -\Delta C + n\Delta s. \tag{8.3}$$

For the change in the portfolio to be stationary, i.e. $\langle \Delta \phi \rangle = 0$, we require

$$n = \left(\frac{\partial C}{\partial s} \right). \tag{8.4}$$

Thus for such a *rational* portfolio we have

$$\phi = -C + s\frac{\partial C}{\partial s}. \tag{8.5}$$

At this point we also introduce the requirement that our portfolio should evolve according to some risk-free asset with interest rate, r_i and obtain

$$\Delta \phi = r_i \phi \Delta t = r_i \left[-C + \frac{\partial C}{\partial s} s \right] \Delta t. \tag{8.6}$$

8.3.2   The theory of option prices

Now consider the log-price return of our asset, $r(t, \delta t)$. We assume this satisfies a Fokker-Planck equation in the style of eqn (7.17), truncated at second order,

$$\frac{\partial P(r, t)}{\partial t} = -\frac{\partial}{\partial r} \left(D_1(r, t)p(r, t) \right) + \frac{\partial^2}{\partial r^2} \left(D_2(r, t)p(r, t) \right). \tag{8.7}$$

In line with Bachelier's approach, eqns (6.12) and (7.5), we chose the coefficient $D_1(x, t)$ to be a constant v_d, the average return for the asset. Fluctuations are assumed to be captured entirely by the second term on the RHS of eqn (8.7).

Thus from Ito's calculus lemma, eqn (7.13), we have

$$\frac{\Delta \langle r \rangle}{\Delta t} = D_1(r, t) = v_d. \tag{8.8}$$

We may use Ito's calculus lemma, eqn (7.13), again to write a similar equation for the price $s = s_0 \exp[r]$,

$$\frac{\Delta \langle s \rangle}{\Delta t} = [D_1(r, t) + D_2(r, t)]s = v_d s + D_2(r, t)s, \tag{8.9}$$

where we used $\partial s / \partial r = \partial^2 s / \partial x^2 = s$.

Finally, using Ito's calculus lemma a third time, we can write an equation for the average or expected option price change over the time period t to $t + \Delta t$,

$$\frac{\Delta \langle C \rangle}{\Delta t} = \frac{\partial C}{\partial t} + D_1(r, t)\frac{\partial C}{\partial r} + D_2(r, \delta t)\frac{\partial^2 C}{\partial r^2}. \tag{8.10}$$

This equation may be rewritten in terms of the asset price, s,

$$\frac{\Delta \langle C \rangle}{\Delta t} = \frac{\partial C}{\partial t} + v_d s \frac{\partial C}{\partial S} + D_2(r,t) \left[s \frac{\partial C}{\partial s} + s^2 \frac{\partial^2 C}{\partial s^2} \right]. \tag{8.11}$$

Making use of eqn (8.9) for the second and third terms on the RHS we obtain

$$\frac{\Delta \langle C \rangle}{\Delta t} - \frac{\Delta \langle s \rangle}{\Delta t} \frac{\partial C}{\partial s} = \frac{\partial C}{\partial t} + D_2(r,t) s^2 \frac{\partial^2 C}{\partial s^2}. \tag{8.12}$$

From eqns (8.3) and (8.4), we now see that the LHS of this equation is the change in value of the portfolio, i.e.

$$-\frac{\Delta \phi}{\Delta t} = \frac{\partial C}{\partial t} + D_2(r,t) s^2 \frac{\partial^2 C}{\partial s^2}. \tag{8.13}$$

We now equate eqns (8.13) and (8.6) to obtain:

$$r_i \left[C - s \frac{\partial C}{\partial s} \right] = \frac{\partial C}{\partial t} + D_2(r,t) s^2 \frac{\partial^2 C}{\partial s^2}. \tag{8.14}$$

8.3.3   The Black-Scholes equation

The celebrated equation of Black and Scholes now follows if we assume that the fluctuations are Gaussian, i.e. $D_2(r,t) = const. = \sigma_0^2/2$, see Chapter 7.3. We then obtain

$$r_i \left[C - s \frac{\partial C}{\partial s} \right] = \frac{\partial C}{\partial t} + \frac{\sigma_0^2}{2} s^2 \frac{\partial^2 C}{\partial s^2}. \tag{8.15}$$

The interesting aspect of this equation is that it is a relatively simple, closed, second-order differential equation for the option price C as a function of the underlying asset price, s. Moreover, the option price depends only on the risk-free interest rate/return, r_i (eqn (8.6); it does not depend on the mean return, v_d, eqn (8.8), of the underlying asset.

Here we present one of what are a number of ways to solve this equation. Note that both the option and asset prices, C and s, are positive. Note also that $0 \le t \le T$. Finally note that at $t = T$ we have the boundary condition, as introduced earlier, eqn (8.1), $C(s,T) = s - K$ for $s > K$ and $C(s,T) = 0$ for $s < K$.

We now proceed by first introducing the change of variable

$$C(s,t) = e^{r_i t} U(s,t). \tag{8.16}$$

Substituting into the Black-Scholes eqn (8.15) and rearranging terms we obtain

$$\frac{\partial U}{\partial t} + r_i s \frac{\partial U}{\partial s} + \frac{\sigma_0^2}{2} s^2 \frac{\partial^2 U}{\partial s^2} = 0. \tag{8.17}$$

Now let $t' = T - t$. (Note the system is evolving backwards from time T) and $r = \ln(s/s_0)$ (log returns should give Brownian motion, possibly with drift) to obtain

$$-\frac{\partial U}{\partial t'} + (r - \sigma_0^2/2) \frac{\partial U}{\partial r} + \sigma_0^2/2 \frac{\partial^2 U}{\partial r^2} = 0. \tag{8.18}$$

Now introduce the further substitutions $r = z - (r_i - \frac{\sigma_0^2}{2})\tau$ and $t' = \tau$. From this we deduce the transformations $\frac{\partial}{\partial z} = \frac{\partial}{\partial r}$ and $\frac{\partial}{\partial \tau} = \frac{\partial}{\partial t'} - (r_i - \frac{\sigma_0^2}{2})\frac{\partial}{\partial r}$, which lead immediately to the final result,

$$\frac{\partial U(z,\tau)}{\partial \tau} = \frac{\sigma_0^2}{2} \frac{\partial^2 U(z,\tau)}{\partial z^2}, \tag{8.19}$$

where $-\infty < z < \infty$ and $0 < \tau < \infty$. This can be recognized as a diffusion equation (see for example eqn (3.44), for the case of zero drift) or heat transfer equation. The solution is given by

$$U(z,\tau) = \frac{1}{2\sqrt{\pi \frac{\sigma_0^2}{2} \tau}} \int_{-\infty}^{\infty} e^{\frac{-(z-z')^2}{4\sigma_0^2/2\tau}} U(z',0)dz'. \tag{8.20}$$

Exercise 8.1 The reader should check that this solution does indeed solve eqn (8.19).

Reintroducing the expression for the variable z in eqn (8.20), writing $r = \ln(s/s_0)$, and using eqn (8.16), the option price may be now written as follows,

$$C(s(\tau),\tau) = e^{-r_i\tau} \frac{1}{2\sqrt{\pi \frac{\sigma_0^2}{2}\tau}} \int_{-\infty}^{\infty} e^{-(z'-\ln s(\tau)-(r_i-\frac{\sigma_0^2}{2}\tau)^2/4\frac{\sigma_0^2}{2}\tau} U(z',0)dz'. \tag{8.21}$$

Note that τ is the time to maturity of the option.

Now introducing the change of variable $S = \ln z'$, we use the initial condition, $U(z',0) = max(s - K, 0)$ and obtain

$$C(s(\tau),\tau) = e^{-r_i\tau} \int_K^{\infty} \frac{e^{-\{\ln s/s(\tau)-(r_i-\frac{\sigma_0^2}{2})\tau\}^2/4\frac{\sigma_0^2}{2}\tau}}{2s\sqrt{\pi \frac{\sigma_0^2}{2}\tau}}(s-K)ds. \tag{8.22}$$

After some manipulation the final result for the value of the option emerges as

$$C = sN(d_1) - Ke^{-r_i\tau}N(d_2), \tag{8.23}$$

where

$$d_1 = \frac{\ln(s/K) + (r_i + \frac{\sigma_0^2}{2})\tau}{\sigma_0\sqrt{\tau/2}}$$

$$d_2 = \frac{\ln(s/K) + (r_i - \frac{\sigma_0^2}{2})\tau}{\sigma_0\sqrt{\tau/2}} = d_1 - \sigma_0\sqrt{\tau/2} \tag{8.24}$$

(see for example Voit (2001)).

The function $N(x)$ is the *cumulative* normal distribution,

$$N(x) = \frac{1}{\sqrt{2\pi}} \int_x^{\infty} e^{-x^2/2}dx = \frac{1}{2}\left[1 + erf(\frac{x}{\sqrt{2}})\right], \tag{8.25}$$

where $erf(x) = \frac{2}{\sqrt{\pi}} \int_0^x e^{-x^2} dx$ is the error function.

Whilst the result, eqn (8.23) looks very complicated, it has an intuitive interpretation when rewritten as follows,

$$C = e^{-r_i \tau} \left(e^{r_i \tau} s N(d_1) - K N(d_2) \right). \qquad (8.26)$$

First note that $N(d_2)$ is essentially the probability that the final stock price will be above K (in other words, that the option will be exercised) so that $K N(d_2)$ is the strike price times the probability that the strike price will be paid. The expression $s N(d_1) \exp(r\tau)$ is the expected value of a variable that equals s if $s > K$ and 0 otherwise. In other words, $e^{r\tau} s N(d_1) - K N(d_2)$ is the expected value of the option at maturity. The above result is therefore just an expression in the manner of our earlier calculation of the price of the call option for ACME Ltd in Section 8.2!

In the 1990s, Myron Scholes together with some other entrepreneurs, notably Robert Merton, set up a new fund management company based around the Black and Scholes theory. 'Long Term Capital Management' (LTCM), as it was called, was initially very successful. However, after approximately two years, things went wrong. Economies in Japan, then Latin America turned sour. A downturn in the US quickly followed and LTCM went bankrupt in spectacular style. The disaster was the largest that had ever been recorded. Trillions of dollars were lost and those involved in LTCM still bear the mental scars. In spite of this, the theory of Black and Scholes is still used widely in the financial community because of its simplicity.

8.4 Implied volatility

The key assumption built into the above approach to computation of option prices is that the volatility is constant, $\sigma^2 = \sigma_0^2/2$. But the volatility in real markets can vary quite markedly. Indeed as we have seen already in Section 2.1.2, Figure 2.1(c), and shall show again in Chapter 9, the fluctuations of asset prices are not Gaussian and do not follow the simple random walk as assumed by Bachelier. So when estimating the volatility from market data for use in the Black and Scholes formulae, some kind of moving average based on historical values of the data must be used. Depending on the time scale, the values then can differ substantially. As a result, the theory is no longer used to compute option prices a priori. Rather, the theory is used to compute so called *implied* volatilities, based on traded option prices quoted in the market place. This implied volatility gives the trader an indication of the level of volatility that one might expect in the future. This might be assumed to be reasonable within the time to maturity of the option.

Using empirically quoted option prices for a variety of expiry dates and strike prices, one can deduce the *market view* for a number of future values of the volatility of an asset. This is referred to as the *term structure of the volatility*. If the Black and Scholes theory were valid, then the term structure of the volatility would yield constant values for different exercise or strike prices. Empirical results show this not to be the case. The results for the implied volatility, when plotted as a function of strike price, K, tend to show a curve in the shape of a smile or grin. Sometimes the curve is symmetric about the value $K = s(t)$; sometimes it is asymmetric. We refer the reader at this point to the many texts written on options for more detail (see, for example, Hull (2006)).

Our key point is that the theory breaks down and, as we noted previously, the outcome can be catastrophic. As we have seen in Section 6.3, (see Figure 6.3(c)), the assumptions leading to the use of Brownian fluctuations greatly underestimate the probability of large fluctuations of stock prices. Other distributions are more appropriate.

8.5 Other developments and issues

There have been many extensions of the theory of Black and Scholes. New kinds of options are used in real markets and numerous books are available within the finance literature that discuss the computation of the value of these various option and derivatives. The core of the theory of Black and Scholes remains. Only the boundary conditions used are different. The problem is essentially a topic in applied mathematics; no new physics is introduced.

A different route was taken by Lisa Borland (Borland, 2002a, 2002b), whose generalization of the Black and Scholes theory deals with non-Gaussian fluctuations by making the coefficient D_2 a function of the probability distribution $p(r, t)$. We will briefly discuss this in Section 10.6.

The present economic crisis is leading many to ask if the concepts of 'risk free rate' and 'risk free assets' are actually meaningful. More fundamentally, the idea that derivatives may be treated as somehow separate from the market has been questioned by physicists Cacciolo, Marsili, and Vivo (2009). In their paper they show that the ideal assumptions that underpin much financial mathematics and engineering are not compatible with market stability. In essence, during the development of the theory of Black and Scholes, the assumption is made that the price of the underlying asset is not affected by the writing of an option contract. This may have been a fair approximation when option contracts were few in amount or small in total value. However it is clear from eqns (8.2) and (8.4) that the person writing an option needs to buy a number, n, of the underlying asset. This imposes a new demand from traders for the asset.

If this demand is small, it may be neglected. However, today the number of options in circulation may be very large. Indeed, there is no reason why the value associated with these options should not exceed the total value of the underlying asset. The associated demand by option traders for the underlying asset may then be significant and its effect on the underlying asset price fluctuations cannot be ignored. We can expect this demand in some way to be proportional to changes in the asset price. Therefore it may accentuate the fluctuations in the manner of a noise trader (see Section 2.2.2), introducing new instabilities into the asset fluctuations. In effect what is required is a greater understanding of the impact of self-transactions on the market. The paper of Caccioli *et al.* (2009) begins to look at this phenomenon, however, a detailed study of this effect is beyond the content of our text. Clearly this is important and needs to be better understood, not only by the scientific community, but also by regulators whose aim is to ensure markets are stable.

9
Asset fluctuations and scaling

Only in Britain could it be thought a defect to be too clever by half. The probability is that too many people are too stupid by three-quarters.
    John Major, Prime Minister UK 1992–7

In the previous chapter we noted the catastrophe that occurred when Merton and Scholes implemented the Black and Scholes theory of option prices. Recall that the basic theory assumes that the log-price fluctuations conform to a Gaussian law. That asset prices did not follow Gaussian statistics had been noticed some years earlier. Indeed Bachelier himself questioned whether real assets actually followed the Gaussian law. His problem was that he had no data that allowed him to properly assess the matter.

Analysing log-price fluctuations for cotton prices using a few hundred data points, Benoit Mandelbrot in 1963 also found that the Gaussian distribution did not fit empirical data. Benoit Mandelbrot, who died recently in 2010, is best known for his work on fractals. However, his 1963 study of asset prices which suggested that asset price fluctuations conform to stable Lévy distributions with fat tails, rather than Gaussian distributions, had profound implications for financial mathematics.

First we discuss the notion of stable distributions and show that the consequence of this assumption is that the characteristic function of the corresponding distribution function has a particular form. In a special limit this yields a Gaussian distribution. More generally, the probability distribution conforms to the Lévy stable function, which has particular scaling properties.

Mantegna and Stanley (1995) studied high frequency (minute-by-minute) asset price data and assessed the degree to which Lévy functions were good fits. For the S&P index from 1984–9, they showed that a Lévy function was a fairly good approximation for short timescales. For larger timescales, the form of the distribution function appeared to change from a Lévy with a power-law tail exponent of about 1.4 to a Gaussian function.

We illustrate the point using the value of the S&P 500 index from 1991–5 and also the Russell 2000 Index from 1997–2001. The conclusions, we shall see, are similar to those obtained with Mantegna and Stanley's S&P data. For small timescales, Lévy stable distribution functions may be a reasonable fit but at longer times (exceeding about 100 minutes), the fluctuations within the measured range seem to evolve into Gaussian fluctuations (see for example the annual data shown in Figure 6.1 and also Figure 10.3).

However at higher values of log-returns, the shape of the distribution supports results published by Gopikrishnan *et al.* (1999) that even the high frequency data does not fit a stable Lévy distribution. Furthermore, returns at short time-scales (less than about five to ten minutes) are not independent and therefore the distribution is no longer stable. For more recent data from the period 2006–10 we find that the computed scaling exponent is outside the permissible range for Lévy distributions, possibly due to this decrease in the decay time.

9.1 Stable distributions

Let us proceed by introducing the concept of *form stable* probability distributions and deriving a particular expression for their characteristic function. A more general expression will be presented in Section 9.2.

In Section 3.2 we saw that we can compute the probability distribution, $p(z)$, of the sum of a set of independent, identically distributed random variables, x_i, by forming the convolution of the elementary probability distributions, $p(x)$. From this, it follows that the corresponding characteristic function for the composite variable, z, is formed by simply multiplying together the elementary characteristic functions for the variables, x_i.

These features are captured by the Chapman-Kolmogorov equation (eqn (4.15)) for Markovian processes when we look for solutions of the form $p(x,t+\delta t|x',t)=p(x-x'|\delta t)=p_{\delta t}(x-x')$. In this case, eqn (4.15) may be re-expressed as a convolution,

$$p_{\delta t_1+\delta t_2}(x'-x) = \int p_{\delta t_2}(x'-x'')p_{\delta t_1}(x''-x)dx''. \tag{9.1}$$

The corresponding Fourier transform is

$$\tilde{p}_{\delta t_1+\delta t_2}(k) = \tilde{p}_{\delta t_1}(k)\tilde{p}_{\delta t_2}(k). \tag{9.2}$$

It follows that the distribution function with respect to the time interval $t = N\delta t$ can be expressed as a generalized convolution of distribution functions corresponding to the time interval δt,

$$\tilde{p}_{N\delta t}(k) = [\tilde{p}_{\delta t}(k)]^N. \tag{9.3}$$

The probability distribution function is said to be *form stable* if the shape of the probability distribution function is independent of the number of convolutions. The idea is expressed formally as

$$p_{t=N\delta t}(x)dx = f(u)du, \tag{9.4}$$

where the variables, x and u are linearly related, viz $u = \alpha_N x + \beta_N$, where α_N and β_N are independent of x and only functions of N.

Using this relationship, we can compute the Fourier transform of the function $f(x)$ as

$$\tilde{f}(k) = \int f(u)e^{iku}du = \int p_{N\delta t}(x)e^{ik[\alpha_N x+\beta_N]}dx = e^{ik\beta_N}\tilde{p}_{N\delta t}(\alpha_N k). \tag{9.5}$$

Imposing the condition of form stability requires this expression to be valid for all values of N. Therefore the following holds,

$$\tilde{p}_{N\delta t}(k) = \tilde{f}(\alpha_N^{-1}k)e^{-i\beta_N\alpha_N^{-1}k}, \tag{9.6}$$

and thus $\tilde{p}_{\delta t}(k) = \tilde{f}(\alpha_1^{-1}k)e^{-i\beta_1\alpha_1^{-1}k}$. Choosing $\alpha_1 = 1$ and $\beta_1 = 0$ we obtain

$$\tilde{f}(\alpha_N^{-1}k)e^{-i\beta_N\alpha_N^{-1}k} = [\tilde{f}(k)]^N. \tag{9.7}$$

Introducing the cumulant generating function, $\Phi(k) = \ln \tilde{f}(k)$ (see eqn (3.20)), the above equation becomes

$$\Phi(\alpha_N^{-1}k) - i\beta_N\alpha_N^{-1}k = N\Phi(k). \tag{9.8}$$

Splitting off the linear component of k, by writing $\Phi(k) = \phi(k) + ick$, where $\phi(k)$ is a function whose properties we specify in the following, and c is a constant, we finally obtain

$$\beta_N = \alpha_N c[\alpha_N^{-1} - N] \tag{9.9}$$

and

$$\phi(\alpha_N^{-1}k) = N\phi(k). \tag{9.10}$$

The expression for β_N gives the shift of the centre of the probability distribution function as a result of the convolution of N steps. We can always put this to zero by a suitable linear change of variables of our basic distribution functions. We are left with the requirement of eqn (9.10) that the function ϕ be *homogeneous*, $\phi(\lambda k) = \lambda^\mu \phi(k)$, where μ is the degree of homogeneity. Assuming α_N is real, we then obtain $\alpha_N = N^{-1/\mu}$. Such a homogeneous form for the function ϕ is satisfied by the following solution:

$$\phi(k) = \begin{cases} c_1|k|^\mu + c_2 k|k|^{\mu-1} & \text{for } \mu \neq 1 \\ c_1|k| + c_2 k \ln|k| & \text{for } \mu = 1, \end{cases} \tag{9.11}$$

where c_1 and c_2 are arbitrary constants. Rescaling the expression for $\mu = 1$ yields $\phi(\lambda k) = \lambda\phi(k) + \lambda c_2 k \ln \lambda$. However, the additional linear term can be included within the shift, β_N.

Choosing $c_2 = 0$, $\mu = 2$ and setting $c_1 = -\sigma^2/2$, we obtain $\phi = -\sigma^2 k^2/2$, which can be recognized as the cumulant generating function for a Gaussian distribution function (see eqn (3.18)). The Gaussian distribution function is then form stable, as can be seen from the complete expression for the Fourier transform that satisfies the relation $\phi(N^{1/2}k) = N\phi(k)$.

Since we know the single step Gaussian distribution function, we can compute the function for an arbitrary number of steps by using eqn (9.4). This, together with the relation $u = \alpha_N x$ where $\alpha_N = (N\delta t)^{-1/2} = t^{-1/2}$ (we set $\beta = 0$ at the outset), yields immediately

$$p_{t=N\delta t}(x) = \frac{\exp(-\frac{u^2}{2\sigma^2})}{\sqrt{2\pi\sigma^2}} \frac{du}{dx} = \frac{\exp(-\frac{(\alpha_N x)^2}{2\sigma^2})}{\sqrt{2\pi\sigma^2}} \alpha_N = \frac{\exp(-\frac{x^2}{2\sigma^2 t})}{\sqrt{2\pi\sigma^2 t}}. \tag{9.12}$$

So again we see that the shape of the function p_t remains a Gaussian under convolution scaled by the factor $\sigma^2 t$ (cf. Section 3.2).

9.2 Lévy distributions and their scaling properties

A more commonly used representation of the characteristic function $\tilde{f}(k)$, dating back to Khintchine and Lévy (1936), is given by

$$\tilde{f}(k) = e^{\Phi(k)} = \tilde{L}_{\alpha,b,m,\mu}(k) = \begin{cases} \exp\left[imk - \alpha|k|^\mu[1 - ib\frac{k}{|k|}\tan(\frac{\pi\mu}{2})]\right] & \text{for } \mu \neq 1 \\ \exp\left[imk - \alpha|k|[1 + ib\frac{2}{\pi}\frac{k}{|k|}\ln|k|]\right] & \text{for } \mu = 1, \end{cases}$$

(9.13)

where α and b are constants, obviously related but not identical to the constants c_1 and c_2 introduced in eqn (9.11). (Note that $\alpha \neq \alpha_N$ and corresponds to the number N of steps). The constant m gives the position of the peak. This expression captures, as we see below, not only Gaussian functions, but a wider range of functions termed Lévy functions.

Detailed analysis shows that the functions so defined conform to probability distributions, i.e. they must be both positive and normalizable, if, and only if, $\alpha > 0$ and the asymmetry parameter b satisfies $|b| \leq 1$.

Leaving aside the drift, function eqn (9.13) represents the characteristic function of a probability density that is form invariant under convolution. The set of such functions are the Lévy functions. The Gaussian distribution is a special subset. Simple expressions for Lévy probability distribution functions can only be calculated in a few special cases: the Gaussian distribution ($\mu = 2, b = 0$), the Lévy-Smirnov distribution ($\mu = 1/2$, b=1), given for the case of a peak at the origin ($m = 0$) by

$$L_{\alpha,1,0,\frac{1}{2}}(x) = \frac{2\alpha}{\sqrt{\pi}(2x)^{3/2}}e^{-\frac{\alpha^2}{2x}} \text{ for } x > 0,$$

(9.14)

and the Cauchy or Lorentz distribution for $\mu = 1, b = 0$, see eqn (3.11).

For large values of the argument, x, the Lévy functions follow a power-law. Consider the symmetric Lévy function ($b = 0$) with a peak at the origin ($m = 0$):

$$L_{\alpha,0,0,\mu}(x) = \frac{1}{\pi}\int_0^\infty \exp(-\alpha|k|^\mu)\cos(kx)dk.$$

(9.15)

This integral can be expressed in a series expansion by expanding the exponential in a power series in $\alpha|k|^\mu$ and evaluating the integrals over k to obtain

$$L_{\alpha,0,0,\mu}(x) = -\frac{1}{\pi}\sum_{n=1}^\infty \frac{(-\alpha)^n}{|x|^{\mu n+1}}\frac{\Gamma(\mu n + 1)}{\Gamma(n + 1)}\sin(\frac{\pi\mu n}{2}),$$

(9.16)

valid for $|x| \to \infty$.

The asymptotic dependence on x is provided from the leading term in the series expansion:

$$L_{\alpha,0,0,\mu}(x) \sim \frac{C}{|x|^{1+\mu}}.$$

(9.17)

The constant $C = \alpha\mu\Gamma(\mu)\sin(\pi\mu/2)$ is positive and is frequently called 'the tail'. It turns out that for the Lévy function to be stable, the exponent μ must lie between 0 and 2; otherwise the Lévy function becomes unstable and ultimately converges to a Gaussian distribution under convolution.

If the Lévy function is not symmetric ($b \neq 0$), then the asymptotic behaviour is given by $L_{\alpha,b,0,\mu}(x) \sim C_+/x^{1+\mu}$ for $x \to +\infty$ and $L_{\alpha,b,0,\mu}(x) \sim C_-/x^{1+\mu}$ for $x \to -\infty$. The asymmetry is characterized by the parameter $b = (C_+ - C_-)/(C_+ - C_-)$. If $1 < \mu < 2$ and the Lévy function is completely asymmetric, ie, $b = 1$, it turns out that the Lévy distribution is a power-law for $x \to \infty$ while it converges to zero as $x \to -\infty$ according to $\exp(-|x|^{\mu/\mu-1})$. The converse occurs if $b = -1$.

Lévy functions with the same exponent, μ and asymmetry coefficient, b are related by the scaling law

$$L_{\alpha,b,0,\mu}(x) = \alpha^{-1/\mu} L_{1,b,0,\mu}(\alpha^{-1/\mu} x). \tag{9.18}$$

The moments can then be written as

$$\langle |x|^\theta \rangle = \int |x|^\theta L_{\alpha,b,0,\mu}(x) dx = \alpha^{\theta/\mu} \int |x|^\theta L_{1,b,0,\mu}(x) dx. \tag{9.19}$$

It follows from the way the Lévy functions decay to zero as $|x| \to \infty$ that these integrals do not always exist. In fact the variance and higher moments are infinite. The absolute value of the spread, i.e. when $\theta = 1$, does however exist. This suggests a characteristic scale of the fluctuations of $\alpha^{1/\mu}$. When $\mu \leq 1$, even the mean and average of the absolute value of the spread diverge. However, the median (see eqn (3.4)) and most probable value (see eqn (3.5)) associated with a Lévy function still exist, even when $\mu \leq 1$.

The cumulative distribution function of the Lévy function conforms to a simple scaling law which can be derived from eqn (9.18), viz

$$C_{\alpha,b,0,\mu}(x) = \int_{-\infty}^{x} L_{\alpha,b,0,\mu}(\xi) d\xi = \alpha^{-1/\mu} \int_{-\infty}^{x} L_{1,b,0,\mu}(\alpha^{-1/\mu}\xi) d\xi$$

$$= \int_{-\infty}^{\alpha^{-1/\mu}x} L_{1,b,0,\mu}(\xi) d\xi = C_{1,b,0,\mu}(\alpha^{-1/\mu}x). \tag{9.20}$$

Alternatively we may write

$$C_{\alpha,b,0,\mu}(\alpha^{1/\mu}x) = C_{1,b,0,\mu}(x). \tag{9.21}$$

We already noted in the introduction that empirical financial data only exhibits Lévy scaling over a finite range of time horizons or returns. Also, while Lévy functions have infinite variance, this is not the case for financial data. One attempt to overcome this problem is to use a *truncated Lévy function*. Such a function, for large values of the returns, takes the form

$$p_{\delta t}(x) \sim C_\pm/x^{1+\mu} \exp(-x/x_0). \tag{9.22}$$

The parameter x_0 is then chosen to accommodate the transition from a Lévy function to a function that converges to zero more rapidly as is evident from the empirical data. However, unlike the motivation to use Gaussian or Lévy functions, the introduction of truncated functions is merely empirical and begs the question as to the origin of the lack of scaling and the basis for using such a truncated function.

We now turn to the examination of empirical financial data and assess to what extent our Lévy scaling results conform to reality.

9.3 Analysis of empirical data

We begin our analysis by looking at 'historic' S&P 500 index data from 1991 to 1995, i.e. we continue the analysis carried out by Mantegna and Stanley for a further five years. The data was obtained from the financial data provider π-Trading.com (http://www.pitrading.com/) and we used the minute closing price, i.e. the price of the last transaction in any minute.

The time series of the closing prices is shown in Figure 9.1(a), with the corresponding log returns shown in Figure 9.1(b). We then compute the probability distribution function of the log price returns, using a procedure as discussed in Section 6.3. Figure 9.1(c) shows these distributions for 1, 3, 9, 27 and 81 minute log-returns. For a Gaussian distribution, such a log-linear plot would result in an inverted parabola, as we had shown in Figure 6.4 for simulated data. Clearly this is not the case here. A question that one might ask next is how do the functions for different time horizons scale? Do they follow the prescription derived above for Lévy functions?

When analysing this issue via the S&P index, Mantegna and Stanley (1995) noted from eqn (9.18) that setting $x = 0$ gives a relation between the peak values of the different Lévy distributions for different time horizons. Using this relation they computed a value of 1.4 for the power-law index μ and used this value to rescale the Lévy functions themselves. However, one can expect such peak values to be sensitive to the binning procedure used in forming the distribution functions from the raw data.

We thus choose a different approach, based on the computation of the cumulative distribution functions for our $N = 1, 3, 9, 27$ and 81 minute log-returns, as shown in Figure 9.2(a). This involves determining the values x_N of log-returns for which the respective cumulative functions take on a specified value C, say $C = 0.5$ in our example. Figure 9.2(b) shows that the ratio x_N/x_1 scales as $N^{1/\mu}$. Here we have identified the exponent from our scaling analysis in Section 9.1, where we have $x_1 = \alpha_N x_N = N^{-1/\mu} x_N$. The curve is independent of the particular values of C that we choose ($C = 0.5, 0.15$ and 0.03 in our example), indicating that the scaling law holds over much of the distribution and we find $\mu = 1.55$ for our S&P 500 data.

Using this value, we can now rescale all the different distribution functions of Figure 9.1(c), so that all our data for the different time-intervals are now described by one functional form, as shown in Figure 9.2(c). (This is often referred to as data *collapse*). The scaling of the data works very well, and the exponent is close to the value of 1.4 that was found by Mantegna by earlier S&P 500 data covering the period January 1984 to December 1989. Mandelbrot's pioneering 1960s analysis of cotton prices resulted in an exponent of $\mu = 1.7$.

An unskewed Lévy distribution with mean zero is fixed by its value at the peak $p(x = 0)$, and the scaling parameter α. Having obtained α from the above figure, we can now attempt a one-parameter fit of the distribution function of our one-minute data. The result is shown in Figure 9.3, and we find that the data is indeed well described by this functional form for log-returns centred around the peak. The fit however over estimates the probability of having large returns, again consistent with the findings of Mantegna and Stanley. In Figure 9.4 we show another successful Lévy scaling analysis, this time for data for the Russell 2000 Index from 1997–2001, resulting in the index $\mu = 1.37$. Curiously, however, our analysis of more recent data does not

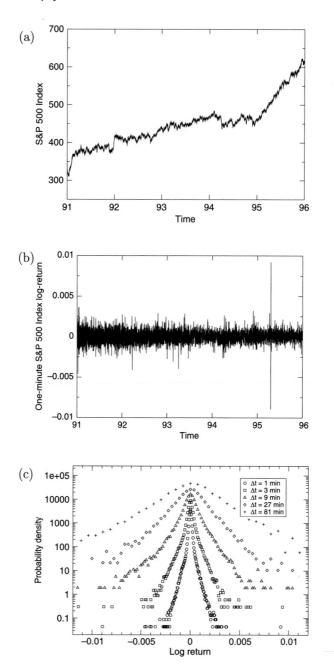

Figure 9.1 Variation of the S&P 500 index from 1 January 1991 to 31 December 1995.
(a) Minute-by-minute data taken from Pi-Trading.com. (b) One-minute log returns. (c) Probability density distribution functions for 1, 3, 9, 27, and 81 minute log returns. The data is shown on a log-linear scale where the deviation from a parabolic form indicates that the distributions are non-Gaussian. For better visual presentation, the distribution functions for 3, 9, 27, and 81 minute data have been vertically shifted.

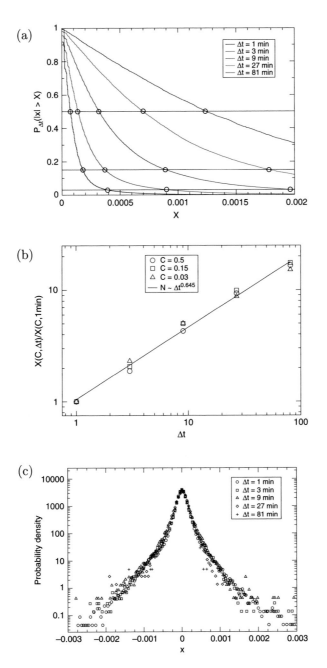

Figure 9.2 Scaling analysis for the S&P 500 log-return data shown in Figure 9.1. (a) Computed cumulative distribution functions for log-return data for $N\Delta t = 1, 3, 9, 21, 81$ minutes. The return values x_N corresponding to the intercepts with the constants $C = 0.5, 0.15$, and 0.03 (solid lines) are used to compute a scaling factor. (b) Values of returns $x(N\text{minutes})/x(1\text{minute})$ for $C = 0.5, 0.15$, and 0.03, as a function of time $N\delta t$, plotted on a log–log scale. The line is obtained from a least square fit, resulting in a slope of $\mu^{-1} = 0.65$. (c) The distribution functions of Figure 9.1(c) have been rescaled according to the prescription of eqn (9.21) with a power-law exponent of $\mu = 1.55$ and are plotted so as to coincide with the one-minute return distribution.

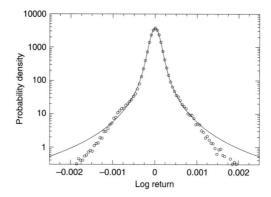

Figure 9.3 Distribution of S&P 500 index minute log-return data from 1991–5, together with a one-parameter fit of the data around the peak to a Lévy distribution with scaling parameter $\mu = 1.55$ (solid line). The fit fails to describe the wings of the data (log-returns $|r| > 0.001$).

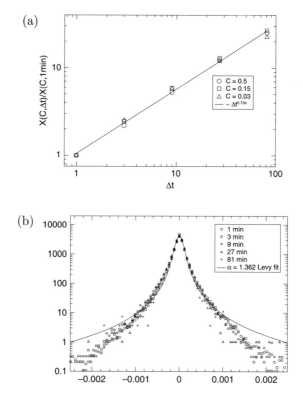

Figure 9.4 Lévy scaling analysis of the Russell 2000 Index for the years 1997–2001 (compare with Figure 9.2). (a) Analysis of the cumulative distribution function results in a scaling exponent $\mu = 1/0.73 = 1.37$. This is used to rescale the data sets for $\Delta t = 1, 3, 9, 27, 81$ minutes and compute the one parameter fit to a Lévy function shown in (b). (Raw data from http://www.pitrading.com.)

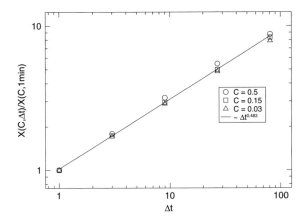

Figure 9.5 A scaling analysis on the Dow Jones index from 2006–10 shows that in this case the scaling exponent $\mu = 1/0.483 = 2.07$. Although close to 2, this is outside the permissible regime of values for Lévy distributions ($1 \leq \mu \leq 2$). A feature confirmed by others. See for example Gopkrishnan *et al.* (1998).

seem to adhere as well to Lévy distributions. This is illustrated in Figure 9.5 for the Dow Jones Index from 2006–2010. While the reason for this is not clear at this stage, it could relate to correlations in the log-return data, and how their typical time scale varies over the years, as we will now discuss.

9.4 Small times and independence

Our discussion so far assumes that the jumps in log returns are independent. From the analysis of daily returns in Section 2.1.3 we know already that the normalized autocorrelation function for the volatility, exhibits correlation out to times of the order of a week or more (see Figure 2.3), whilst the jumps in the log-returns appear to be independent, see Figure 2.2.

In Figure 9.6(a) we plot the autocorrelation function of the one-minute log-returns of our S&P 500 data from 1991–95 of Section 9.3 as a function of time lag on a log-lin scale. The roughly linear decrease points to an exponential decay and we can determine a decay time of about three minutes. This is in line with the analysis of S&P 500 data from 1984–96 by Stanley and colleagues (Liu *et al.* 1999) who found a decay constant of about four minutes.

As the frequency of trading changes with advancing computer technology, it is of interest to examine how this decay constant changes with time. In Figure 9.6(b) we plot the autocorrelation function of one-minute log-returns as computed from S&P 500 data for the six time intervals 1983–9, 1987–90, 1991–4, 1999–2002, 2003–6 and 2007–11. Fits to exponentials result in a decay time of about six minutes in 1983–9 down to less than 0.3 minutes in 2007–11. This decrease in decay time is approximately exponential, as shown in Figure 9.6(c).

We must conclude that financial time series are not only not form-stable, but also that the high frequency log-returns are not independent.

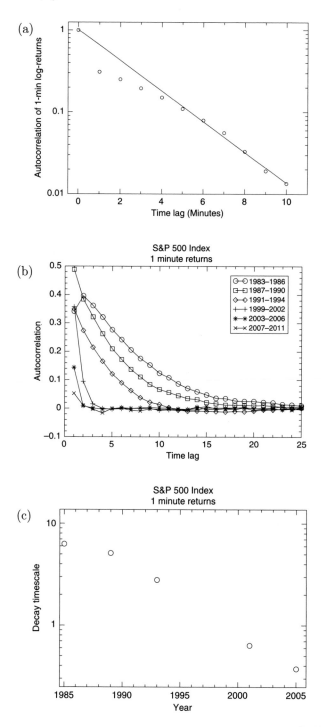

Figure 9.6 Autocorrelations in one-minute return data for the S&P 500 index. (a) A log-linear plot of the normalized autocorrelation function for one-minute S&P 500 returns from 1991–5 shows a roughly linear decay, with a decay time of approximately three minutes. (b) Data for six-year time intervals shows a steady, and approximately exponential, decrease of this decay time (c).

9.5 The Heston model

In 1993, Heston proposed a stochastic model in which not only the log-price return, but also the volatility, is allowed to fluctuate in a stochastic manner (Heston, 1993). Thus the log-price returns (after subtracting the drift) follow the familiar stochastic equation

$$dr = -\frac{v}{2}dt + \sigma dW^{(1)}, \tag{9.23}$$

(cf. Section 7.7, eqn (7.44)) and the variance $v = \sigma^2$ is modelled as

$$dv = \gamma(v_0 - v)dt + \kappa\sqrt{v}dW^{(2)}. \tag{9.24}$$

Here, v_0 is the long-term mean of the variance v and γ is the relaxation rate of this mean, which in the previous sections was assumed to be extremely large, relative to the fluctuation rate of the log-returns themselves. $W^{(1)}$ and $W^{(2)}$ are two standard Wiener processes, that are correlated as

$$dW^{(2)} = \rho dW^{(1)} + \sqrt{1 - \rho^2}dZ, \tag{9.25}$$

where dZ is another Wiener process, independent of $dW^{(1)}$. Together they satisfy a new and more general Fokker-Planck equation

$$\frac{\partial P}{\partial t} = \gamma\frac{\partial}{\partial v}[(v - v_0)P] + \frac{1}{2}\frac{\partial}{\partial x}(vP) + \rho\kappa\frac{\partial^2}{\partial x\partial v}(vP) + \frac{1}{2}\frac{\partial^2}{\partial x^2}(vP) + \frac{\kappa^2}{2}\frac{\partial^2}{\partial v^2}(vP), \tag{9.26}$$

where $P = P(r, v|v_i)$ is the transition probability to have a log-return r, and variance v, at time t, given variance v_i at time $t = 0$ and initial log-return zero.

An analytical solution was obtained by Drăgulescu and Yakovenko (2002), which enabled computation of the unconditional distribution function $P(x)$, using the expression

$$P(r) = \int_0^\infty dv_i \int_0^\infty dv P(r, v|v_i)\Pi^*(v_i), \tag{9.27}$$

where $P(v_i)$ is the stationary distribution of the initial variance v_i. In the long time limit, the solution exhibits scaling behaviour, $P \propto K_1(z)/z$. Here $K_1(z)$ is a first order modified Bessel function and z is given by $z = \sqrt{\tilde{x}^2 + \bar{t}}$, where $\tilde{x}$ and $\bar{t}$ are scaled values for the log-price returns and time, respectively.

Fitting the few parameters in the model, Silva and Yakovenko (2003) applied this result to stock price fluctuations for the US Dow Jones and S&P indices for the period 1982–99. They obtained good agreement (over many decades of magnitude) for times greater than one day, through to 250 days. However, deviations appear once the data for 2000 and more recent dates are considered. The authors recognise that this arises as a result of assuming a mean growth rate which is constant. During the period 2000–13 is immediately clear this is a poor assumption as the mean growth rate has moved from positive to negative and back again; it has been far from constant.

A more sophisticated approach which, at the very least, can model the time dependent behaviour of this exogenous parameter, is required. To date, such a theory has not been published. Anticipating our discussions in Chapters 15 and 20, we note that the basic analysis of fluctuations based on the Heston model corresponds to admitting a fluctuating 'inverse dither' within the method of super-statistics. More studies based on this method might be interesting to evaluate using this alternate approach.

10
Models of asset fluctuations

ZUFALL, ich weiß, das Wort hat keinen Inhalt. Zufall läßt sich im voraus berechnen, nur daß die Rechnung sich sehr in die Länge zieht, und das Resultat sagt dann nicht mehr, als daß die Wahrscheinlichkeit eines Zufalls fast unwahrscheinlich ist, aber auch launisch. Das Unwahrscheinliche kann also jedem doch jederzeit zugemutet werden, das wäre dann Zufall. Ich weiß, das Wort hat keinen Inhalt.

CHANCE, I know this word is lacking in content. Chance may be computed in advance, but the calculation is very lengthy, and the result then simply states that the probability of pure chance is nearly improbable, but also moody. Thus one is expected to put up with the improbable at any time, that would then be chance. I know, the word is lacking in content.

Reinhard Baumgart, Hausmusik - Ein deutsches Familienalbum.

From the previous chapter we concluded that asset price fluctuations are not form stable, and the characterization by Lévy functions is an approximation that does not describe too well the extreme fluctuations in financial time series. In this chapter we construct (semi-) analytical models of asset fluctuations using the framework of Fokker-Planck equations, together with the generalised diffusion constants first introduced in Section 7.1.

In Section 10.1 we show that simple trial forms for D_1 and D_2 result in a distribution function of returns that has a power law tail. However, while the model captures the exponential decay of the linear autocorrelation function of returns, it fails to reproduce the long range correlations exhibited by square returns (Section 10.2).

In Section 10.3 we generalize the approach using a trial *time dependent* function for the density distribution. This provides a more successful route to the characterization of the long- and short-time nature of both autocovariance functions, as is demonstrated for empirical financial data in Section 10.4.

Finally, using the Fokker-Plank equations, we show in Section 10.6 how this approach, that accounts for the fat tails in the distribution functions, may be applied to the computation of option prices.

10.1 Generalized diffusion coefficients and the distribution function of returns

Let us begin by recalling the definition of the generalized n-th order diffusion coefficients $D_n(x,t)$ from eqn (7.3),

$$D_n(x,t) = \frac{1}{n!} \lim_{\Delta t \to 0} \int \Delta t^{-1} (x'-x)^n P(x', t+\Delta t | x, t) dx'$$

$$= \frac{1}{n!} \lim_{\Delta t \to 0} \frac{\langle [x(t+\Delta t) - x(t)]^n \rangle_{x(t)=x}}{\Delta t}. \tag{10.1}$$

Here we choose the random variable $x(t)$ to be the log-price return, i.e. $r(t, \delta t) = \ln \frac{s(t+\delta t)}{s(t)}$, making the D_n a function also of δt, as will be discussed below.

If we can compute these diffusion coefficients we can then deduce the probability distribution function from the Fokker-Planck eqn (7.17) and the various correlation functions from eqn (7.11).

Limiting the terms in the sum on the RHS of the Fokker-Planck equation to the first two, the probability distribution function for log-returns is now given by

$$\frac{\partial p(x, t)}{\partial t} = -\frac{\partial}{\partial x}[D_1(x, t)p(x, t)] + \frac{\partial^2}{\partial x^2}[D_2(x, t)p(x, t)]. \tag{10.2}$$

In the approximation chosen by Bachelier and by Einstein, D_2 is set to a constant, independent of x, $D_2 = D$ ($= k_BT/\zeta$ for the case of Brownian motion, see eqn (7.7)), and D_1 is a constant drift velocity, $D_1 = v_d$.

Choosing D_1 to remain independent of time, but allowing for an x-dependence, $D_1 = D_1(x)$, we can introduce the *potential function* $V(x)$ defined via the relation

$$-\frac{\partial V(x)}{\partial x} = \frac{k_BT}{D}D_1(x). \tag{10.3}$$

The Fokker-Planck eqn (10.2) then reduces to the familiar diffusion equation with drift governed by D_1 (see eqn (3.44)) and yields a stationary solution, p^*, equal to a Boltzmann distribution,

$$p^*(x) = \frac{1}{Z}\exp(-V(x)/k_BT) = \frac{1}{Z}\exp\left[\int^x \frac{D_1(x')}{D}dx'\right], \tag{10.4}$$

where Z is a normalization factor.

Exercise 10.1 The reader may wish to show that p^* is indeed a solution of the diffusion equation with drift, eqn (3.44).

We can see that the above choices for the diffusion coefficients lead to fluctuations that are Gaussian distributed. However, from our previous discussion we know asset fluctuations are not Gaussian and can have so-called 'fat tails'. So how can this be achieved within the framework of Fokker-Plank equations?

In the more general case, when the diffusion coefficient D_2 is no longer a constant, but a function of x only, we may obtain a stationary solution

$$p^*(x) = \frac{\exp\left[\int^x \frac{D_1(x')}{D_2(x')}dx'\right]}{ZD_2(x)}, \tag{10.5}$$

which may be rewritten as

$$p^*(x) = \frac{\exp\left[-\left(\ln(D_2(x)) - \int^x \frac{D_1(x')}{D_2(x')}dx'\right)\right]}{Z}. \tag{10.6}$$

We can formally identify the potential function $W(x)$ (sometimes referred to as the *objective function*), using the expression

$$p^* = \exp(-W(x))/Z. \tag{10.7}$$

Thus we obtain

$$W(x) = \ln(D_2(x)) - \int^x \frac{D_1(x')}{D_2(x')} dx'. \tag{10.8}$$

At this point, specific forms for the functions D_1 and D_2 are required in order to progress. Let us for simplicity of our argument set

$$D_1(x) = -\kappa d D_2(x)/dx, \tag{10.9}$$

where κ is a constant. (We shall in a moment see that this assumption is justified, based on an analysis of empirical stock data). This allows us to compute the integral in eqn (10.8) to obtain

$$W(x) = (1 + \kappa) \ln D_2(x), \tag{10.10}$$

and the probability distribution function

$$p^*(x) = \frac{\exp[-(1 + \kappa) \ln(D_2(x))]}{Z} = \frac{1}{Z[D_2(x)]^{1+\kappa}}. \tag{10.11}$$

At this stage we need specific forms for D_1 and D_2, in order to progress. The simplest choices are linear and quadratic variations with log-return x and this is indeed supported by empirical data, as follows.

Both diffusion coefficients may be directly computed from stock data, where we have for the following analysis used minute-by-minute data from the Dow Jones index over the period 1993–2012. Using the definition of log-returns we obtain

$$D_n(x, \delta t) = \frac{1}{n!} \lim_{\Delta t \to 0} \frac{\langle [x(t + \Delta t, \delta t) - x(t, \delta t)]^n \rangle_{x(t,\delta t)=x}}{\Delta t}$$
$$= \frac{1}{n!} \lim_{\Delta t \to 0} \frac{\langle [x(t + \delta t, \Delta t) - x(t, \Delta t)]^n \rangle_{x(t,\delta t)=x}}{\Delta t} \tag{10.12}$$

where the conditional ensemble average is now evaluated as a time average, as follows.

A time-window of width δt is moved along the times series of our one-minute returns to obtain a sequence of $N - \delta t$ return differences $(x(t + \delta t, 1\text{min}) - x(t, 1\text{min}))^n$, where N is the number of returns in our time series (approx. 1.7 million). The data is then binned according to the value of x at the beginning of the window. From this sequence we obtain the average difference of return, for the given window width as a function of x.

Figure 10.1 shows the evaluation of both D_1 and D_2 for our minute-by-minute Dow Jones index log-returns. The data is well described by

$$D_1(x, 1\text{min}) = -2\kappa DCx \tag{10.13}$$

and

$$D_2(x, 1\text{min}) = D(1 + Cx^2) \tag{10.14}$$

where D and C are constants. Note that these forms are consistent with our previous assumption, eqn (10.9). Such linear and quadratic relationships for D_1 and D_2 have also been shown to hold for daily data, see Queiros *et al.* (2007).

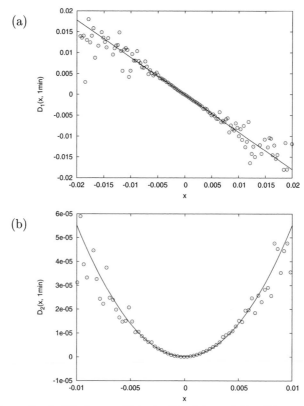

Figure 10.1 Generalized diffusion coefficients D_1, (a), and D_2, (b), computed from minute-by-minute Dow Jones data over the period 1993–2012. Lines are least-square fits to the data and suggest that D_1 is linear in the x coordinate (i.e. the log-return r) and D_2 is quadratic. Raw Dow Jones data (over 1.7 million points) taken from the Tickwrite 6 database (http://www.tickdata.com).

Using these expressions for D_1 and D_2, the probability distribution now reduces to

$$p^*(x) = \frac{1}{\tilde{Z}[1 + Cx^2]^{1+\kappa}}, \tag{10.15}$$

where the constant D has been absorbed into the normalization factor $\tilde{Z} = ZD^{1+\kappa}$.

Exercise 10.2 The reader might wish to show that $\tilde{Z}$ is given by

$$\tilde{Z} = C^{1/2} \frac{\Gamma(1+\kappa)}{\Gamma(1/2)\Gamma(\kappa + 1/2)}. \tag{10.16}$$

The approach presented in this section thus leads to a probability distribution of returns with a power-law tail, as is required to describe financial data. However, in order to account for the key features of such data, we also need to capture the existing correlations in the returns. As we shall see in the following section, this is not fully achieved with the current approach.

10.2   Correlation functions

In Section 7.2 we had established that the temporal evolution of the average of a function $M = M(x,t)$ of a random variable, x, may be determined in terms of an expansion involving the n-th order diffusion constants, see eqn (7.11).

Choosing $M = x$ and considering only the first and second diffusion constants, eqn (7.11) gives

$$\frac{d \langle x \rangle_{x(t=0)=x_0}}{dt} = \langle D_1(x,t) \rangle. \tag{10.17}$$

Note that in the above equation and in what follows we have used t to refer to the time-lag, previously called δt, in order to simplify the notation.

Choosing $M = x^2$ gives

$$\frac{d \langle x^2 \rangle}{dt} = 2 \langle x D_1(x,t) \rangle + 2 \langle D_2(x,t) \rangle. \tag{10.18}$$

Inserting the simple expression used previously for D_1 and D_2 (eqns (10.13) and (10.14)), leads to a closed pair of equations for the correlation functions. For the linear correlations we have

$$\frac{d \langle x \rangle}{dt} = -2\kappa DC \langle x \rangle, \tag{10.19}$$

yielding

$$\langle x(t) \rangle = x_0 e^{-2\kappa DCt}. \tag{10.20}$$

In the above we have dropped the subscripts $x(t=0) = x_0$ for brevity of expression. $\langle x(t) \rangle$ corresponds to the un-normalized correlation function $G(t)$ (see eqn (2.11)), i.e.

$$G(t) = \langle x(0)x(0+t) \rangle = x_0 \langle x(t) \rangle_{x(t=0)=x_0}. \tag{10.21}$$

The linear correlations are predicted to decay exponentially with time. This describes well our minute-by-minute Dow Jones data, as shown in Figure 10.2 which displays the normalized autocorrelation function $R(t)$, eqn (2.12). The decay constant is given by $b \equiv 2\kappa DC = 1.92$ min^{-1}, corresponding to a relaxation time of approximately 0.52 minutes. This value may be compared with that obtained by directly computing the generalized diffusion coefficient, $D_1(x,t)=-bx$ (from eqn (10.12), see Figure 10.1(a)) where we find $b = 0.89$ min^{-1}, corresponding to 1.12 minutes for the relaxation time.

Stanley has published calculations using the Dow Jones index that suggest a decay constant of about 10–15 minutes and thus higher than the value we obtain here. So perhaps the exponential decay constant is not universal. Indeed in the nineteenth century, stocks were traded more on the scale of weeks rather than minutes so the decay constant in those times was longer than today (Sabatelli *et al.* 2002).

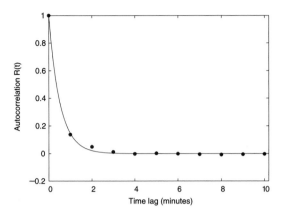

Figure 10.2 Normalized autocorrelation function $R(t) = \langle x(t)\rangle/x_0$ for minute-by-minute Dow Jones data over the period 1993–2012. The data is well described by an exponential decay, as suggested by eqn (10.20) with a decay constant of about 31 seconds (0.52 minutes).

Let us now turn to the square correlations. According to our hypothesis for the diffusion coefficients they satisfy the relation

$$\frac{d\langle x^2\rangle}{dt} = -4\kappa DC\langle x^2\rangle + 2D\langle 1 + Cx^2\rangle$$
$$= -2DC(2\kappa - 1)\langle x^2\rangle + 2D. \tag{10.22}$$

Integrating this equation yields

$$\langle x^2(t)\rangle = \left(x_0^2 - \frac{1}{C(2\kappa - 1)}\right)\exp[-2DC(2\kappa - 1)t] + \frac{1}{C(2\kappa - 1)}. \tag{10.23}$$

Exercise 10.3 The reader might wish to derive this result.

So this approximation predicts, providing $\kappa > 1/2$, that the decay of the square correlations is also exponential. Furthermore, it is readily shown that providing $\kappa < 1$ the square correlations decay more slowly than the linear autocorrelations.

Whilst for small time, the decay of the square correlations for financial data might be exponential and decay more slowly than that for the linear correlations, we have already noted in Section 2.1.3 that for very long times the square correlations exhibit a slow power-law-like decay (see also Figure 2.3). This is not predicted from our theoretical approach presented in this section. We will thus now approach the problem in a slightly different manner by making both D_1 and D_2 time dependent, in order to capture the long-time correlations seen in empirical data.

10.3   Time dependent distribution function and scaling

Recall in Section 3.2 we derived the distribution function, $p(x)$, for Gaussian fluctuations, eqn (3.43). For the case of zero drift this may be expressed as

$$p(x) = \frac{1}{Z} \exp(-\beta x^2). \tag{10.24}$$

Since the distribution function is normalized, it follows that $\beta Z^2 = \pi$. If we allow β and Z to be functions of time, then provided the functions $Z(t)$ and $\beta(t)$ scale correctly with time, $p(x,t)$ will describe the fluctuations for the different times. The correct scaling is obtained by choosing the function to be a solution of the diffusion equation (eqn (3.44), with $v_d = 0$),

$$\frac{\partial p(x,t)}{\partial t} = \frac{\partial^2}{\partial x^2} D(t) p(x,t) \tag{10.25}$$

After a little algebra one obtains the requirements

$$\frac{d\beta(t)}{dt} = -4\beta(t)^2 D(t) \quad \text{and} \quad D(t) = -\frac{1}{4\beta(t)^2} \frac{d\beta(t)}{dt}. \tag{10.26}$$

For the simple diffusive process of Bachelier and Einstein where $D(t) = D =$ constant we recover the familiar result (cf. eqn (3.43))

$$p(x,t) = \sqrt{\frac{1}{4\pi Dt}} \exp\left(-\frac{x^2}{4Dt}\right). \tag{10.27}$$

Exercise 10.4 The reader may wish to carry out the above steps.

However, from our previous investigation we know that the Gaussian distribution is a poor description of asset price fluctuation data. A better fit to the data for a specific time lag t is given by the function

$$p(x,t) = \frac{1}{Z(t)} \left[1 + \frac{\beta(t)x^2}{\alpha}\right]^{-\alpha}, \tag{10.28}$$

as will be demonstrated in Section 10.4.

As in the example above, $Z(t)$ and $\beta(t)$ are two functions via which we try to capture the time dependence of $p(x,t)$ and α is a parameter that we shall insist later to be larger than $3/2$. The nature of this distribution is that in the limit $\beta x^2/\alpha \to 0$ it reduces to a Gaussian distribution function (or Boltzmann distribution function, see eqn (10.4)) when β may be interpreted as an 'inverse temperature'.

Exercise 10.5 The reader may wish to show this.

It is readily shown that the normalization function, $Z(t)$, or in the language of statistical physics, the *partition function*, and time dependent inverse temperature $\beta(t)$ are linked via the relation

$$Z(t) = \int dx \left[1 + \frac{\beta(t)x^2}{\alpha}\right]^{-\alpha} = \sqrt{\frac{\alpha}{\beta(t)}} B(\frac{1}{2}, \alpha - \frac{1}{2}), \tag{10.29}$$

where $B(x,y) = \Gamma(x)\Gamma(y)/\Gamma(x+y)$ is the standard Beta function or Euler integral of the first kind.

Exercise 10.6 The reader may wish to show this by making use of the normalisation condition $\int_{-\infty}^{+\infty} p(x,t)dx = 1$.

Similarly one can show that

$$\sigma^2(t) = \langle x^2(t)\rangle = \int dx p(x,t)x^2$$
$$= \frac{\alpha}{\beta(t)} \frac{\int \frac{z^2 dz}{(1+z^2)^\alpha}}{\int \frac{dz}{(1+z^2)^\alpha}} = \begin{cases} \infty \text{ if } \alpha < 3/2 \\ \frac{\alpha}{(2\alpha-3)\beta(t)} \text{ if } \alpha > 3/2. \end{cases} \tag{10.30}$$

Exercise 10.7 The reader may wish to show this expression for the variance, keeping in mind that $\langle x \rangle = 0$, as was set out at the beginning of this chapter, in Section 10.1.

Let us continue to assume, given the result of our empirical analysis (Figure 10.1), that D_1 is linear in x, i.e.

$$D_1(x,t) = -b(t)x, \tag{10.31}$$

but now admit a time dependent coefficient $b(t)$ which we shall later assume, in line with our empirical analysis, to vary approximately inversely with time lag, t. Introducing a time dependent function for the generalised diffusion coefficient, D_1, implies that the autocorrelation function for linear returns no longer decays as a simple exponential (cf. eqn (10.17)) and we return to this point at the end of Section 10.5.

We shall also assume that $D_2(x,t)$ is a quadratic function of x,

$$D_2(x,t) = D(t)\left(1 + \frac{\beta(t)x^2}{\alpha}\right) \tag{10.32}$$

but now with the addition of a time dependence, included in the term $D(t)$ of yet unspecified form.

At this state we will proceed as in Section 10.1 and set $D_1(x,t) = -\kappa\frac{dD_2(x,t)}{dx}$, where κ is a constant. This immediately results in

$$b(t) = 2\kappa\frac{D(t)\beta(t)}{\alpha}, \tag{10.33}$$

consistent with our empirical data, as shown in Section 10.4, resulting in $\kappa = 1$.

Substituting the resulting expressions for D_1 and D_2 into the Fokker-Planck eqn (10.2) and using the probability distribution of eqn (10.28) we obtain, after a little algebra, the following,

$$\frac{1}{Z(t)}\frac{dZ(t)}{dt} = -\frac{1}{2\beta(t)}\frac{d\beta(t)}{dt} = -2\left(\kappa + (1-\alpha)\right)\frac{D(t)\beta(t)}{\alpha}. \tag{10.34}$$

Exercise 10.8 The reader may like to carry out the somewhat lengthy derivation of the above result.

In order to proceed, let us assume that $D(t)$ is dependent only on $\beta(t)$ raised to some power γ, viz:

$$D(t) = D(\beta(t)) = D_0 \beta^\gamma(t), \tag{10.35}$$

where D_0 is constant, leading to $b(t) = \frac{2\kappa D_0}{\alpha} \beta^{\gamma+1}$. Again this scaling ansatz will be justified later by our empirical data, see Section 10.4.

Using the ansatz we obtain the solution

$$\beta(t)^{-(\gamma+1)} = \beta_0^{-(\gamma+1)} + \frac{4D_0(\alpha - 1 - \kappa)}{\alpha}(\gamma+1)t, \tag{10.36}$$

leading to the scaling

$$\beta(t) \propto t^{-\frac{1}{\gamma+1}} \quad \text{and hence} \quad D(t) \propto t^{-\frac{\gamma}{\gamma+1}} \quad \text{and} \quad b(t) \propto t^{-1} \tag{10.37}$$

in the limit $t \to \infty$.

Let us now see how the above results compare with empirical deductions from our Dow Jones data.

10.4 Comparison with financial data

Figure 10.3 (a) shows the probability distribution function for one-minute log-price returns for the Dow Jones over the period 1993–2012. In line with our analysis above we have removed the mean (which a very small correction for the data considered here), so that the distribution is centred around $x = 0$. The data is well described by the proposed probability function of eqn (10.28), with power law exponent $\alpha = 1.84\pm0.04$. The tail exponent is thus $-2\alpha \simeq 3.7$, similar to values obtained for other datasets (see our discussion in Section 6.4). Since the power law tail is most pronounced in the one-minute data, we keep it fixed in all the fits of log-price returns for different time windows, i.e. 10, 100 and 400 minutes, as shown in Figures 10.3 (b), (c) and (d).

Initial values of $Z(t)$ are obtained from reading off $P(0,t) = 1/Z(t)$ for the various time windows. We then perform a one free parameter fit to eqn (10.28) to obtain an initial guess for $\beta(t)$, followed by a two parameter fit to obtain final values for $Z(t)$ and $\beta(t)$. The resulting curve for $\beta(t)$ is shown in Figure 10.4 in a double-logarithmic plot. The data is best described by two power-law regimes. For small times ($t <\sim 10$ min) β varies as $t^{-1.34}$, for larger times it varies as $t^{-1.}$. Correspondingly we obtain from eqn (10.37) that for times up to about 10 minutes $\gamma \simeq -1/4$ while for larger time we have $\gamma = 0$.

Since the exponent $\alpha = 1.84$ of the probability distribution of log-returns exceeds $3/2$, we see from eqn (10.30) that $\sigma^2(t) \sim \beta^{-1}(t)$. For small times we thus obtain $\sigma^2(t) \sim t^{1.34}$, corresponding to a *super-diffusive* regime, while for larger times $\sigma^2(t) \sim t$, and normal diffusion prevails.

We have also computed the generalised diffusion constants $D_1(x,t)$ and $D_2(x,t)$ for our Dow Jones (1993–2012) dataset, using the functional form of eqns (10.31) and

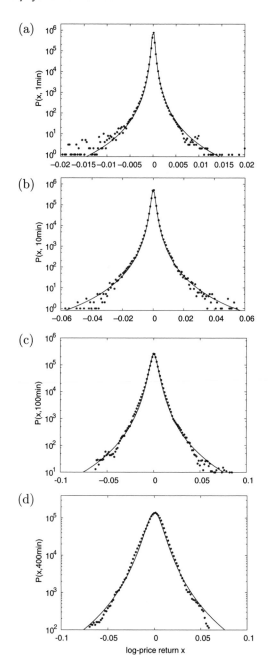

Figure 10.3 Plots of the probability density distribution for log-price returns for time lags 1, 10, 100, and 400 minutes. The transition from a distribution with what appears to be a cusp to a Gaussian shape as time increases is clearly visible. The solid lines are fits to the probability function of eqn (10.28), resulting in an exponent $\alpha = 1.84 \pm 0.04$. (Dow Jones minute-by-minute data, 1993–2012)

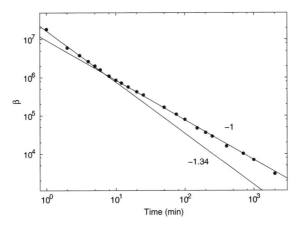

Figure 10.4 Variation of β as a function of the time lag. The data is obtained from fits of the Dow Jones minute-by-minute data to the distribution function of eqn (10.28) as shown in Figure 10.3. For times less than about ten minutes the data is well described by $\beta(t) \propto t^{-1.34}$ while for larger times $\beta(t)$ varies as t^{-1}. (Data shown on double logarithmic scale.)

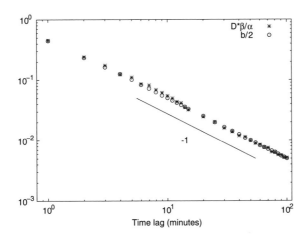

Figure 10.5 Empirical values for $b(t)/2$ obtained from fitting the generalised diffusion constants $D_1(x,t)$ to eqn (10.31). The data is well described by a power law with exponent -1. Also shown in this graph is the variation of the fitted prefactor of the quadratic in $D_2(x,t)$, i.e. $D(t)\beta(t)/\alpha$, see eqn (10.32). This graph thus justifies eqn (10.33) with $b(t)$ varying as $b(t) = 0.89/t$.

(10.32), respectively. Figure 10.5 shows that the pre-factor $b(t)$ is well described by $b(t) \propto t^{-1}$ over the range from 1 to 100 minutes. For larger time lags our data proved to be too noisy to allow for a meaningful comparison of b and β.

From fits to our data for $D_2(x,t)$ to eqn (10.32), we obtained the prefactor to the quadratic in x, i.e. $D(t)\beta(t)/\alpha$. Values of this fitted prefactor as a function of time are shown in Figure 10.5. We see that the values scale like $b(t)$ and we can indeed overlay $b(t)$ and $\kappa D(t)\beta(t)/\alpha$ for a value of $\kappa = 1$, justifying eqn (10.33).

Let us return to the time interval from one to ten minutes, where β varies as $t^{-1.34}$ and thus $\gamma \simeq -0.25$. Interestingly we see that this corresponds to $\gamma \simeq -1/2\alpha$. In this regime we may then choose

$$D(t) = D_0 Z^{1/\alpha}(t). \tag{10.38}$$

This conjecture is consistent with the Fokker-Plank equation for the fluctuations being equivalent to a non-linear equation,

$$\frac{\partial P}{\partial t} = -\frac{\partial}{\partial x} D_1(x,t)P + D_0 \frac{\partial^2}{\partial x^2} P^{1-1/\alpha} \tag{10.39}$$

This equation has been proposed by Tsallis in the context of non-extensive statistical mechanics (see Gell-Mann and Tsallis, 2004, and also Section 20.4). By varying the parameter α, the equation has been found to fit probability distribution functions for a wide range of systems from science and economics.

10.5   Volatility correlation function revisited

Now let's return and look again at eqn (10.18) for the correlation function $\langle x^2 \rangle$ using the generalized diffusion coefficients defined by eqns (10.31) and (10.32) respectively. We have

$$\frac{d\langle x^2(t)\rangle_{x_0}}{dt} = 2D(t)\left(1 + \frac{\beta(t)}{\alpha}(1 - 2\kappa)\langle x^2(t)\rangle\right). \tag{10.40}$$

Assuming the power law expressions (eqn (10.37)) for $D(t)$ and $\beta(t)$ we have:

$$\frac{d\langle x^2(t)\rangle_{x_0}}{dt} = 2D_0\beta(t)^\gamma\left(1 + \frac{\beta(t)}{\alpha}(1 - 2\kappa)\langle x^2(t)\rangle\right). \tag{10.41}$$

Let us consider the regime $t > 10$ minutes, where $\gamma = 0$ and $\beta(t) = \beta_2 t^{-1}$, with $\beta_2 = const$.
This results is

$$\frac{d\langle x^2(t)\rangle_{x_0}}{dt} = 2D_0\left(1 + \frac{\beta_2}{\alpha}(1 - 2\kappa)\langle x^2(t)\rangle t^{-1}\right). \tag{10.42}$$

The solution of this equation is given by

$$\langle x^2(t)\rangle_{x_0} = \frac{G_{x^2}(t)}{x_0^2} = \frac{2D_0}{1 - 2D_0\frac{\beta_2}{\alpha}(1 - 2\kappa)} t + const. \times t^{2D_0\frac{\beta_2}{\alpha}(1-2\kappa)}. \tag{10.43}$$

But now we recall that the correlation function we actually require is $R_{x^2}(t)$, i.e. $R(t)$ of eqn (2.14), applied to x^2. Noting that the non-conditional average $\langle x^2(t)\rangle^2 \simeq G_{x^2}(t)$ in the limit of $t \to \infty$, where the first term of the RHS of eqn (10.43) is the dominant term, we obtain

$$R_{x^2}(t) \simeq const. \times t^{2D_0\frac{\beta_2}{\alpha}(1-2\kappa)}. \tag{10.44}$$

Recalling that $\kappa \simeq 1$ we find that $R_{x^2}(t)$ scales as $t^{-2D_0\beta_2/\alpha}$.

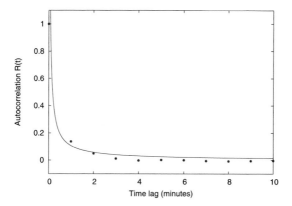

Figure 10.6 One-parameter fit of a power-law with exponent $b_1 = 0.89$ to the normalized autocorrelation function $R(t) = \langle x(t) \rangle / x_0$. While the fit is inferior to the exponential fit shown in Figure 10.2, and diverges at zero time lag, it arises from a theory that captures well the long time behaviour of the volatility of log-returns. (Minute-by-minute Dow Jones data over the period 1993–2012).

The ratio $2D_0\beta_2/\alpha$ is obtained from our fits for $D_2(x,t)$, as it is twice the prefactor of the x^2 term at $t = 1$ minute. We finally obtain

$$R_{x^2}(t) \propto t^{-0.9}, \tag{10.45}$$

in good agreement with our data, as shown in Figure 2.3.

While our theoretical approach is successful in delivering a power law for the volatility of log-returns for large times, we now recall our introduction in Section 10.3 of a time dependent function $D_1(x,t)$, see eqn (10.31). The time dependence of $b(t)$, $b(t) = b_1/t$, leads immediately to a power law for the auto-correlation function,

$$\langle x \rangle / x_0 = (t/t_0)^{-b_1}. \tag{10.46}$$

Figure 10.6 shows a one-parameter fit of this power law to our data, using the value $b_1 = 0.89$, as obtained from our data for $b(t)$, see Figure 10.5. We see that this approximation provides a not unreasonable empirical fit to the data. However, unlike an exponential decay, which is bounded as t tends to zero, the power law diverges for short times. For very small times, the simple model presented here may break down and $b(t)$ might no longer scale inversely with time. Indeed some might be surprised that this model fits as well as it does over the entire time period of our data, given the limited number of parameters involved. Further generalisations to include, for example, skewness in the data might be considered for future work.

Today high frequency trading is the norm, making it much harder to profit from trends in the autocorrelation function for linear returns. Nevertheless, with high performance computing, rapid analysis of these interconnected characteristics of financial data is possible (extended also to include order book analysis, which is beyond the scope of this text), and should be used more widely by anyone who is seriously interested in the behaviour of financial markets.

10.6   Non-Gaussian fluctuations and option pricing

The approach to non-Gaussian fluctuations as introduced above may be used to obtain a new perspective on option prices, as discussed in Chapter 8. Previously in our treatment (Section 8.3.3) we assumed that the diffusion coefficient D_2 was constant. However, taking our lead from the above discussion it is possible to suppose that D_2 is both a function of x and t, see eqn (10.32).

Borland (Borland, 2002a, 2002b) has proposed such a generalization of the Black and Scholes theory, choosing D_2 to have a non-linear dependence on the distribution function, $p(x, t)$, viz

$$D_2(x, t) = Dp^q(x, t). \tag{10.47}$$

$D_1(x, t)$ is assumed to be unchanged and equal to a constant average return v_d for the asset, as before (see eqn (8.8)). The function p then satisfies a generalized non-linear Fokker-Plank equation which is simply derived from eqn (8.7), viz:

$$\frac{\partial p(x, t)}{\partial t} = -\frac{\partial}{\partial x} v_d p(x, t) + D \frac{\partial^2}{\partial x^2} p^{1+q}(x, t) \tag{10.48}$$

The equation for the option price is then no longer given by eqn (8.14), but is now

$$r_i \left[C - s \frac{\partial C}{\partial s} \right] = \frac{\partial C}{\partial t} + p^q(x, t) D s^2 \frac{\partial^2 C}{\partial s^2}. \tag{10.49}$$

Eqns (10.48) and (10.49) together must be solved self-consistently to obtain the option price. This is a more complicated model of option prices. Furthermore we now see that both the interest rate, r_i, of the risk free asset and the mean rate of return, v_d, of the underlying asset are included as model parameters. More detail is given in the papers by Borland (Borland, 2002a, 2002b).

11
Risk

Take risks: if you win you will be happy; if you lose you will be wise.
Anon

Risk is normally associated with extreme events. While extreme events can take
asset prices in a positive as well as a negative direction, risk generally refers to negative
or adverse outcomes. In this chapter we shall first introduce the statistics associated
with extreme events and then go on to discuss how these ideas have been used to
construct portfolios where the risk is (hopefully) minimized.

11.1 Statistics of extreme events

Consider a series of fluctuations $x_1, x_2, \ldots x_N$ occurring over time, as indicated in
Figure 11.1. These may be considered to be N realizations of the random variable x
and we may ask the question: what is the probability P that the maximum value,
x_{max}, within this set of N values is less than a specified value Λ?

For simplicity, consider these realisations to be independent and identically distributed (i.i.d) and therefore arising from a single probability distribution, $p(x)$. It
follows that the probability $P(x_{max} < \Lambda)$ is given by

$$P(x_{max} < \Lambda) = p(x_1 < \Lambda)\, p(x_2 < \Lambda) \ldots p(x_N < \Lambda) = C_<(\Lambda)^N = [1 - C_>(\Lambda)]^N,$$
(11.1)

where $C_<(\Lambda)$ and $C_>(\Lambda)$ are the cumulative and complementary cumulative distribution function, as defined in eqns (3.2) and (3.3), respectively.

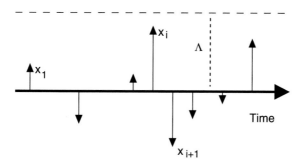

Figure 11.1 Fluctuations x_i as a function of time. Aim is to compute the probability that
their maximum value lies below a specified threshold value Λ.

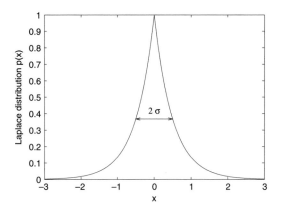

Figure 11.2 The Laplace distribution function of eqn (11.4) has the properties $p(0) = (2\sigma)^{-1}$ and $p(\sigma) = p(0)/e$. (In the graph σ is set to 1.)

If, consistent with our assumption of extreme events, Λ is large, then clearly $C_>(\Lambda)$ is small and much less than unity. The term in brackets may then be approximated by an exponential. Thus

$$P(x_{max} < \Lambda) \approx \exp[-NC_>(\Lambda)]. \tag{11.2}$$

Suppose, to be specific, that we require to be 50% confident that $x_{max} < \Lambda \equiv \Lambda_{1/2}$, or in other words that $P(x_{max} < \Lambda_{1/2}) = 1/2$. Then it follows that $C_>(\Lambda_{1/2}) = \frac{\ln 2}{N}$.

We of course generally require confidence to an arbitrary level Λ_s. That is, the risk is $100s\%$. Now we have $P(x_{max} < \Lambda_s)$, hence

$$C_>(\Lambda_s) = \int_{\Lambda_s}^{\infty} dx p(x) = -\frac{\ln s}{N}. \tag{11.3}$$

This result is, apart from the assumption of independent fluctuations, essentially exact. But now we see that the precise value of Λ_s depends on the shape of the distribution function, $p(x)$. First consider the Laplace distribution,

$$p(x) = \frac{1}{2\sigma} e^{-x/\sigma}, \tag{11.4}$$

where the parameter σ characterizes the width of the distribution, as illustrated in Figure 11.2.

Substituting this distribution into eqn (11.3) we obtain

$$-\frac{\ln s}{N} = \frac{1}{2\sigma} \int_{\Lambda_s}^{\infty} dx e^{-x/\sigma} = \frac{e^{-\Lambda_s/\sigma}}{2}. \tag{11.5}$$

Hence

$$\Lambda_s = \sigma[\ln(1/2) - \ln(\ln s^{-1}) + \ln N] \xrightarrow[N\to\infty]{} \sigma \ln N. \tag{11.6}$$

Clearly in the limit of $N \to \infty$ the maximum value of Λ_s and hence the risk is dominated by the dependence on the width, σ, of the distribution function and the variation with N is relatively small.

Repeating this calculation for a Gaussian distribution function,

$$p(x) = \frac{e^{-x^2/2\sigma^2}}{\sqrt{2\pi}\sigma}, \tag{11.7}$$

we obtain [1]

$$-\frac{\ln s}{N} = C_>(\Lambda_s) = \frac{1}{2}\mathrm{erfc}(\frac{\Lambda_s}{\sqrt{2}\sigma}) \sim \exp(-\Lambda_s^2/2\sigma^2)\frac{\sigma}{\sqrt{2\pi}\Lambda_s}(1 - \frac{\sigma^2}{\Lambda_s^2} + ...). \tag{11.8}$$

This yields

$$\Lambda_s \sim \sqrt{2}\sigma[\ln N - \ln(-\ln s) + \cdots]^{1/2} \xrightarrow[N\to\infty]{} \sqrt{2}\sigma[\ln N]^{1/2}. \tag{11.9}$$

Again the form of the outcome is essentially dependent on the distribution width, σ and the variation with N is relatively small.

However, consider the case where the distribution function has a power-law 'fat' tail for $x > 0$, i.e.

$$p(x) \sim \mu A/x^{1+\mu}, \tag{11.10}$$

where μ and A are constants. This results in

$$-\frac{\ln s}{N} = C_>(\Lambda_s) = \mu A \int_{\Lambda_s}^{\infty} \frac{dx}{x^{1+\mu}} = \frac{A}{\Lambda_s^\mu}, \tag{11.11}$$

leading to

$$\Lambda_s = \left(\frac{AN}{\ln(1/s)}\right)^{1/\mu}. \tag{11.12}$$

Now the leading term in the expression for Λ_s depends on both N and s via the power law exponent μ. This is altogether a more complex relationship and more sensitive to changes in N than in the previous expressions for the exponential and Gaussian distributions.

We have illustrated this in Table 11.1. For power law distributions, the maximum fluctuation can increase by an order of magnitude, as N changes by two orders of magnitude as opposed to fluctuations arising from exponential or Gaussian distribution functions, where the maximum fluctuation barely changes with N. For exponential and Gaussian return distributions, the width of the distribution has often been used as a proxy for risk and this is exploited in the development of the 'efficient portfolio' discussed later. For power-law distribution functions such a simple characterization of risk is not possible.

In our earlier discussion on diffusion (Section 3.2, eqn (3.42)) we derived that the width of the Gaussian distribution is proportional to the square root of time, $\sigma(t) = \sigma_0\sqrt{t}$. Since the width is a measure of the standard deviation of the time

---

[1]Here we have used the definition of the complementary error function, $\mathrm{erfc}(z) = \frac{2}{\sqrt{\pi}} \int_z^{\infty} e^{-t^2} dt$,

together with the first two terms of its asymptotic expansion ($z \to \infty$), $\mathrm{erfc}(z) \simeq \frac{e^{-z^2}}{z\sqrt{\pi}}(1 - \frac{1}{2z^2})$ (Abramowitz and Stegun, 1965).

Table 11.1 Illustration of the dependence of the value of Λ_s on the number of independent events, N for the three types of probability distributions discussed in the text (eqns (11.6), (11.9) and (11.12), respectively).

Distribution	N	500	5000	50000
Laplace	$\ln N$	6.21	8.52	10.82
Gaussian	$\sqrt{\ln N}$	2.49	2.92	3.29
Power-law, $\mu = 3/2$	$N^{2/3}$	23.00	292.40	1357.21
Power-law, $\mu = 3$	$N^{1/3}$	7.94	17.10	36.84

series at time t, we may thus ask the question: is there a time T beyond which the contribution to the change in log-price return from the drift $v_d t$ is larger than that from stochastic fluctuations. This latter contribution is determined by the width, $\sigma(t)$ of the Gaussian distribution function, so we can estimate T as

$$v_d T = \sigma_0 \sqrt{T}, \text{ thus } T = (\sigma_0/v_d)^2. \tag{11.13}$$

The situation may be illustrated roughly by looking at the Unilever share price over the period 1996 to 2012, as shown in Figure 11.3(a). Unilever appears to behave as a stock for which an investment will be ultimately dominated by the drift and make money for the investor in the long term. Even if one buys at the top of the market, the graph suggests that by holding on sufficiently long, the holder of the stock ought to be in profit as the drift grows linearly in time, while the growth of the fluctuations is with the square root in time.

However, one might begin to worry about applying this simple approach when looking at the graph for Vodafone over the same period, see Figure 11.3(b). The fluctuations in this example do not seem to fit, even approximately, the picture of a simple Gaussian, superimposed on a trend. The peak occurred during the dot-com boom at the beginning of the twenty-first century and it is difficult to imagine this stock dynamic being controlled by simple diffusive processes.

There are other problems associated with the application of the theory. We have assumed that the limit $N \to \infty$ may be taken. However, even if we assume a correlation time of 15 minutes which is in line with the results we obtained in Section 9.4, then over a trading time of one year one only has around 6000 statistically independent price increments. Over a period of ten years this amounts to only around 60000 such data points. It is debatable whether this may be considered a large number consistent with the limit $N \to \infty$.

Leaving all these issues aside, it is instructive to develop the ideas of the *efficient portfolio*, first proposed by the economist Harry Markowitz in the 1950s (Markowitz, 1959).

11.2 The efficient portfolio

The simplest case, first considered by Markowitz, is to consider an investor who wishes to know the optimum way to invest in a portfolio of N uncorrelated assets, which follow Gaussian statistics.

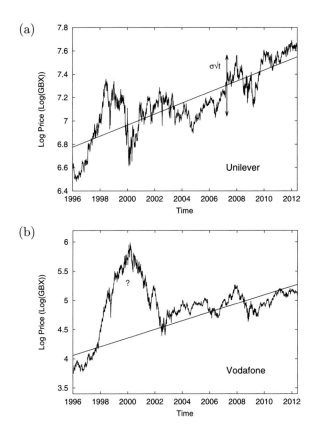

Figure 11.3 The evolution of the logarithm of the share price for the period 1996 to 2012 for (a), Unilever, and (b), Vodafone shares. In the case of Unilever shares it seems that the drift outweighs the contribution from fluctuations in the long term. However, the standard model of Gaussian noise cannot account for the log price variation of Vodafone shares.

Assume each asset i has a price s_i and the number of such assets in the portfolio is n_i. The portfolio value is then $\phi = \sum_i n_i s_i$. The total number of assets is $N = \sum_i n_i$. It is convenient to introduce the portfolio value per asset $\Pi = \phi/N = \sum_i p_i s_i$ where the fraction of assets per category $p_i = n_i/N$ and $\sum p_i = 1$.

The total dividend D from the portfolio is given in terms of the dividend per share, D_i and yield (%) per share, r_i, for each asset i. Since $D_i = r_i s_i$ it follows that $D = \sum_i D_i n_i = \sum_i r_i s_i n_i$.

The return $\bar{r}$ on the total portfolio is then

$$\bar{r} = D/\phi = \sum_i \frac{n_i s_i}{\phi} r_i. \tag{11.14}$$

It is convenient to introduce $q_i = n_i s_i/\Phi$ where

$$\sum_i q_i = 1. \tag{11.15}$$

Thus

$$\bar{r} = \sum_i q_i r_i. \tag{11.16}$$

For uncorrelated Gaussian assets we now introduce the risk,

$$\sigma^2 = \sum_i \sigma_i^2 q_i^2, \tag{11.17}$$

where the σ_i^2 are the variances of the different assets.

The optimal portfolio is now given by minimizing the risk σ^2, given the return $\bar{r}$, defined by eqn (11.16), and the additional constraint of eqn (11.15). This is achieved using the method of *Lagrange multipliers*, as is introduced in Appendix 11.7.

We thus form the Lagrangian $\mathcal{L}$,

$$\mathcal{L} = \sigma^2 - \lambda\bar{r} - \beta\sum_i q_i, \tag{11.18}$$

where λ and β are Lagrange multipliers. Thus

$$\frac{\delta\mathcal{L}}{\delta q_i} = \frac{\delta}{\delta q_i}\left(\sum_i q_i^2\sigma_i^2 - \lambda\sum q_i r_i - \beta\right) = 0. \tag{11.19}$$

This is solved by

$$2q_i\sigma_i^2 - \lambda r_i - \beta = 0. \tag{11.20}$$

We proceed by assuming that one of the assets is riskless, say, $\sigma_0 = 0$, with return r_0. It follows that

$$\beta = -\lambda r_0, \quad \text{and} \quad 2q_i\sigma_i^2 = \lambda(r_i - r_0). \tag{11.21}$$

Hence

$$q_0 = 1 - \sum_{i=1}^{N} q_i = 1 - \lambda\sum_{i=1}^{N}\frac{r_i - r_o}{2\sigma_i^2},$$

$$\text{and } q_i = \frac{\lambda(r_i - r_0)}{2\sigma_i^2} \quad \text{for } i = 1...N. \tag{11.22}$$

Inserting eqn (11.22) into the constraint $\bar{r} = q_0 r_0 + \sum_{i=1}^{N} q_i r_i$, we now obtain λ as

$$\lambda = \frac{2(\bar{r} - r_0)}{\sum_{i=1}^{N}\frac{(r_i - r_0)^2}{\sigma_i^2}} \tag{11.23}$$

and

$$\sigma^2 = \sum_{i=1}^{N} q_i^2\sigma_i^2 = \lambda^2\sum_{i=1}^{N}\frac{(r_i - r_0)^2}{4\sigma_i^2} = \frac{(\bar{r} - r_0)^2}{\sum_{i=1}^{N}\frac{(r_i - r_0)^2}{\sigma_i^2}}. \tag{11.24}$$

We see that the risk σ^2 is a quadratic function of the deviation of the portfolio return from that of the riskless asset, $(\bar{r} - r_0)^2$. According to this theory, we can expect a specific return by accepting a specific risk. Increasing our expected return comes at the expense of increasing risk.

As we will see in the next section, assets are generally correlated, and these correlations should be taken into account when computing optimum portfolios. We will introduce two techniques whose applications originate from physics, namely minimum spanning trees (Section 11.4) and an analysis based on random matrices (Section 11.5).

11.3 Portfolios and correlated assets

Temporal correlations between the log-returns for stocks i and j may be described by a correlation (or covariance) matrix, as was introduced in eqn (3.24). Normalization by the variance of the respective log-returns leads to correlation coefficients ρ_{ij} that lie between $-1 \leq \rho_{ij} \leq 1$, where -1 corresponds to completely anti-correlated, +1 to completely correlated and 0 to uncorrelated values.

These coefficients are given by

$$\rho_{i,j} = \frac{\langle r_i r_j \rangle - \langle r_i \rangle \langle r_j \rangle}{\sqrt{(\langle r_i^2 \rangle - \langle r_i \rangle^2)(\langle r_j^2 \rangle - \langle r_j \rangle^2)}}, \qquad (11.25)$$

with averages

$$\langle r_i \rangle = \frac{1}{T} \sum_{n=1}^{T} r_i(t_n), \quad \langle r_i^2 \rangle = \frac{1}{T} \sum_{n=1}^{T} r_i^2(t_n),$$

$$\langle r_i r_j \rangle = \frac{1}{T} \sum_{n=1}^{T} r_i(t_n) r_j(t_n). \qquad (11.26)$$

Here $r_i(t_n) = \ln s_i(t_n) - \ln s_i(t_{n-1})$ with $s_i(t_n)$ the price of stock i at time t_n. The t_n are equally spaced, referring for example to daily data, and T is the number of records in the time series that is being analysed.

By defining normalized returns $\tilde{r}_i$ as

$$\tilde{r}_i = \frac{r_i - \langle r_i \rangle}{\sqrt{\langle r_i^2 \rangle - \langle r_j \rangle^2}}, \qquad (11.27)$$

we may re-write the correlation coefficients as

$$\rho_{ij} = \langle \tilde{r}_i \tilde{r}_j \rangle. \qquad (11.28)$$

The coefficients for N stocks form a symmetric N by N matrix C with diagonal elements equal to unity that can be represented as

$$C = \frac{1}{T} \langle GG^t \rangle, \qquad (11.29)$$

where G is an N by T matrix with elements $\tilde{r}_i$, and G^t denotes the transpose of G.

Figure 11.4 shows the time evolution of the log-price $\ln(s(t))$ for four stocks from the London FTSE 100 over the period January 2003 to January 2012. The discontinuity

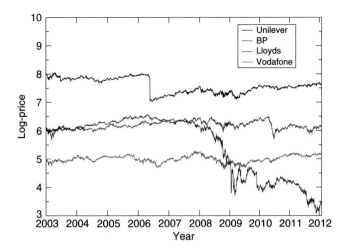

Figure 11.4 Variation of the log-price for four FTSE 100 traded stocks over the period January 2003 through to January 2012 (daily data from yahoo.com).

in the price of Unilever is due to a stock split, when shares were consolidated two for one. The fall in Lloyds illustrates the beginning of the financial crisis in 2008. The Gulf of Mexico oil spill in mid 2010 is responsible for the drop in BP at that time. Using this data we can use eqn (11.28) to compute the correlation function for pairs of stocks.

Figure 11.5 shows this for the pair of stocks, Lloyds-BP. The graph shows the value of $\rho_{\text{Lloyds-BP}}$ computed as a moving average, where the average is taken over a 100 day moving window (T=100). For much of the time window covered by this data (January 2003 through to January 2012) the correlation coefficient is small, suggesting the stocks are uncorrelated. However, there are periods where the correlation coefficient rises significantly. These correspond to the aftermath of the 2000 dot-com boom, the onset of the 2008 financial crisis where FTSE 100 stocks as a whole became much more correlated, and to mid 2011, as the global financial turmoil continues.

11.4 Portfolio analysis using minimum spanning trees

Mantegna and Stanley (2000) have proposed a simple and convenient way to visualize correlations between a large number of stocks via a minimum spanning tree (MST) or an indexed hierarchical tree, i.e. by using concepts developed in graph theory.

In order to compute these objects it is first necessary to introduce a quantity which can be thought of as *distance*. To this end we will introduce a T-dimensional vector $\tilde{\mathbf{r}}_i$ whose components are the normalized log-returns $\tilde{r}_{i,n} = \tilde{r}_i(t_n)$ with $1 \leq n \leq T$, see eqn (11.27).

A distance d_{ij} between vectors $\tilde{\mathbf{r}}_i$ and $\tilde{\mathbf{r}}_j$ (thus corresponding to stocks i and j) may then be defined using the Euclidean norm as follows,

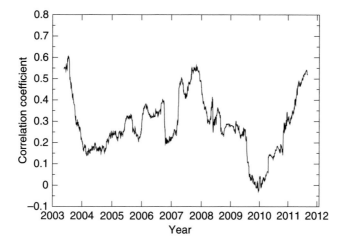

Figure 11.5 A plot of the correlation coefficient $\rho_{\text{Lloyds-BP}}$ for Lloyds and BP for the data shown in Figure 11.4. The coefficient is computed as a moving average with a time window of 100 days (daily data, obtained from yahoo.com). Periods of high correlations may be linked to specific external events.

$$d_{ij} = ||\tilde{\mathbf{r}}_i - \tilde{\mathbf{r}}_j|| = \left(\sum_{n=1}^{T} (\tilde{r}_{i,n} - \tilde{r}_{j,n})^2 \right)^{1/2}. \tag{11.30}$$

The distance may be expressed in terms of the correlation coefficients ρ_{ij} (see eqn 11.28),

$$d_{ij}^2 = (\tilde{\mathbf{r}}_i - \tilde{\mathbf{r}}_j) \cdot (\tilde{\mathbf{r}}_i - \tilde{\mathbf{r}}_j) = ||\tilde{\mathbf{r}}_i||^2 + ||\tilde{\mathbf{r}}_j||^2 - 2\tilde{\mathbf{r}}_i \cdot \tilde{\mathbf{r}}_j = 2(1 - \rho_{ij}), \tag{11.31}$$

where we have made use of the normalization $||\tilde{\mathbf{r}}_i||^2 = ||\tilde{\mathbf{r}}_i||^2 = 1$ (see eqn 11.27).

Since $-1 < \rho_{ij} < 1$, it follows that $0 < d_{ij} < 2$. Since $d_{ij} = 0 \Leftrightarrow i = j$ and $d_{ij} = d_{ji}$, and also the so-called 'triangular' inequality holds, i.e. $d_{ij} \leq d_{ik} + d_{kj}$, d_{ij} satisfies all the properties associated with a distance.

The matrix of distances between pairs of stocks i and j may now be used to create a network of nodes with each node representing a time series of a stock. Stocks that are highly correlated then correspond to nodes that are close together on the network. A *minimum spanning tree* (MST) is a network consisting only of $N-1$ links out of the total of $N(N-1)/2$ possible available links between N different stocks. The $N-1$ links are chosen so that the total length of the network is minimized, under the conditions that all nodes are connected and that there are no loops. (The absence of loops turns the network into a *tree* in the language of graph theory.)

Various algorithms are available to compute an MST (which is unique only if all distance d_{ij} differ). *Prim's algorithm* is given as follows.

- Determine the element with the minimum distance of the matrix and connect the corresponding nodes i and j.
- Determine the minimum distance amongst the remaining matrix entries.

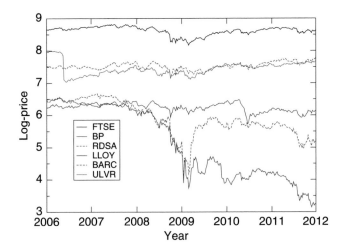

Figure 11.6 The log-price of five FTSE 100 stocks traded at the London Stock Exchange, together with the value of the FTSE itself, over the period 2006–12 (weekly data from http://www.yahoo.com). (BP – British Petrol, RDSA – Royal Dutch Shell plc, LLOY – Lloyds Banking Group plc, BARC – Barclays plc, ULVR – Unilever)

- Connect the corresponding nodes, provided only one of them is already connected to other nodes.
- Repeat this procedure until all nodes are connected.

In this way the minimum spanning tree may be fully constructed.

Let us illustrate this with a rather simple, specific example with a portfolio of six stocks from the London Stock Exchange, see Figure 11.6. Weekly data over the period January 2006 to January 2012 has been used to first compute the correlation matrix, ρ_{ij}, (see Table 11.2) and then the distance matrix, d_{ij} (see Table 11.3).

The minimum distance is between the pair LLOY and BARC (0.46), so these will be connected. The next minimum distance is between BP and RDSA (0.56), but none of these are already connected, so the next pair, BP and BARC (1.16) needs to be considered, and will be connected. This is now followed by the links BP and RDSA (0.56) and RSDA and ULVR (1.56), which completes the procedure, resulting in the minimum spanning tree as shown in Figure 11.7(a). In Figure 11.7(b) we show the MST for all the data of Table 11.3; the FTSE index is now in the centre.

An alternative representation of the correlation between stocks is in the form of an *indexed hierarchical tree* which may be constructed from the minimum spanning tree, as described in the book by Mantegna and Stanley (2000).

Minimal spanning trees have been used to explore the correlations and clustering in stock portfolios in the various markets across the globe. Stocks traded on a market are often classified according to economic sectors, such as Resources, or Basic Industries, with various subdivisions in these sectors. Coelho *et al.* (2007a) have used a minimum spanning tree analysis to examine the clustering of stocks traded on the FTSE 100. By doing so, they could show that some stocks were better reclassified from their present sector. A new classification scheme, adopted by FTSE in the beginning of 2006, offered

Table 11.2 Computed correlation matrix for the 5 FTSE 100 stocks and the value of the FTSE from Figure 11.6. (Note that this is a symmetric matrix, with entries of 1.00 (full correlation) as diagonal elements.)

	FTSE	BP	RDSA	LLOY	BARC	ULVR
FTSE	1.00	0.72	0.81	0.58	0.62	0.29
BP		1.00	0.72	0.32	0.42	0.18
RDSA			1.00	0.34	0.38	0.22
LLOY				1.00	0.77	0.11
BARC					1.00	0.15
ULVR						1.00

Table 11.3 Computed distance matrix, obtained from the correlation matrix of Table 11.2. (Note that this is a symmetric matrix, with entries of 0.00 (distance zero) as diagonal elements.)

	FTSE	BP	RDSA	LLOY	BARC	ULVR
FTSE	0.00	0.75	0.62	0.92	0.87	1.19
BP		0.00	0.75	1.17	1.08	1.28
RDSA			0.00	1.15	1.11	1.25
LLOY				0.00	0.68	1.33
BARC					0.00	1.30
ULVR						0.00

an improvement for the descriptions of the correlations seen in the data. However, still not all correlations were captured appropriately, which could have affected the design of optimum portfolios at the time.

Coelho *et al.* (2007b) also used an MST analysis to study the process of market integration for a large group of national stock market indices, as shown in Figure 11.8. Using a moving time window, they showed how the asset tree evolves over time and described the dynamics of its normalized length, mean occupation layer, and other parameters. Over the period studied, 1997–2006, the minimum spanning tree shows a tendency to become more compact, implying that global equity markets are increasingly interrelated. The consequence for global investors is a potential reduction of the benefits of international portfolio diversification.

The use of minimum spanning trees has been criticized for involving only a subset of the complete set of elements of the correlation matrix. One could imagine using a different subset, corresponding, for example, to the maximal spanning tree. Would this have any relevance? It is not so straightforward to address this question. For this reason we turn now to a different method that exploits the complete set of elements of the correlation matrix in a more systematic manner.

11.5 Portfolio analysis using random matrix theory

With real physical systems, it is usually possible to relate correlations to underlying interactions between basic sub-units such as atoms or molecules. However, this is

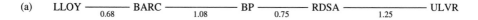

(a)   LLOY ——— BARC ——— BP ——— RDSA ——— ULVR
            0.68   1.08   0.75   1.25

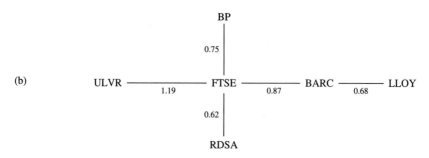

Figure 11.7 Minimal spanning trees constructed from the distance matrix of Table 11.3 for FTSE 100 data over the period 2006–12. (a) MST for the BP, RDSA, LLOY, BARC, and ULVR stocks. (b) MST that also includes the data for the correlation of these stocks with the FTSE index itself.

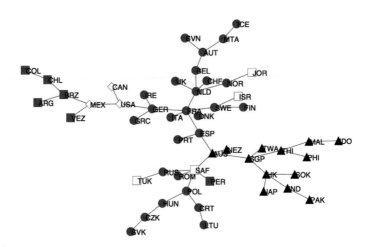

Figure 11.8 Average minimum spanning tree constructed from data for markets of 53 countries over the period 1997–2006. Coding is: Europe, grey circles; North America, white diamonds; South America, grey squares; Asian-Pacific area, black triangles; and 'other' (Israel, Jordan, Turkey, South Africa), white squares. (Figure reprinted from Coelho *et al.* (2007b), with permission from Elsevier.)

not possible for stocks or financial assets, where the underlying interactions are not known. More difficulties arise when analysing the significance and meaning of cross-correlations determined empirically. Firstly, market conditions change with time and correlations may not be stationary. Secondly, the finite length of time series used to estimate the correlations can introduce 'measurement noise'. But it is pertinent to ask the question: can we estimate, from a correlation matrix, those stocks that on average remain correlated over any particular time period? One answer lies in comparing the statistics of the empirical correlation matrix with those of a random correlation matrix and assessing the deviations. To develop this approach we first need to understand a little about random matrices.

11.5.1 Elements of random matrix theory

The methods behind random matrix theory (RMT) have a rich history and were originally developed more than 50 years ago by Wigner, Dyson, and others to explain the statistics associated with energy levels in complex nuclei. They proposed that the Hamiltonian H describing a complex heavy nucleus was made up of independent random elements H_{ij}, drawn from a probability distribution. So for a physical system, the RMT predictions were an average over all the possible interactions. Deviations from these predictions then provided insights into the system specific or non-random properties of the empirical system being considered.

Of particular interest is the distribution of eigenvalues of the matrix under consideration. It can be shown that for a real symmetric matrix with independent and identically distributed elements with finite variance σ^2 of the distribution, the density $\rho(\lambda)$ of the eigenvalues λ is given by

$$\rho(\lambda) = \frac{1}{2\pi\sigma^2}\sqrt{4\sigma^2 - \lambda^2} \quad \text{if } 4\sigma^2 \geq \lambda^2, \tag{11.32}$$

else ρ is zero (see Appendix 11.8). This is the 'semicircle' law first derived by Wigner.

In our case, we see from the previous section that we have an $N \times T$ matrix G composed of N time series each with T elements. The matrix is generally not square but we construct a square $N \times N$ correlation matrix by computing $C = (1/N)GG^t$.

Provided G is square, we can use Wigner's semicircle law to obtain, from the properties of G, the corresponding properties of the matrix C. Thus the eigenvalues of C are obtained from those of G by squaring them: $\lambda_C = \lambda_G^2$. Furthermore, the density of eigenvalues of C is related to that of the density of eigenvalues of G via $\rho(\lambda_C)d\lambda_C = 2\rho(\lambda_G)d\lambda_G$, where the factor 2 takes into account that there are two solutions, $\lambda_G = \pm\sqrt{\lambda_C}$. This yields

$$\rho(\lambda_C) = \frac{1}{2\pi\sigma^2}\sqrt{\frac{4\sigma^2 - \lambda_C}{\lambda_C}} \tag{11.33}$$

This density is valid for $4\sigma^2 > \lambda_C$; else it is zero.

For the case when G is not square, ie $N \neq T$, it can be shown that in the limit $T \rightarrow \infty$ and keeping the ratio $Q = T/N$ fixed, the spectrum of eigenvalues, λ_C, of the square matrix C is bounded, and distributed according to the density

$$\rho(\lambda_C) = \frac{Q}{2\pi\sigma^2}\frac{\sqrt{(\lambda_{\max} - \lambda_C)(\lambda_C - \lambda_{\min})}}{\lambda_C}, \tag{11.34}$$

where $\lambda_{min} < \lambda < \lambda_{max}$ and

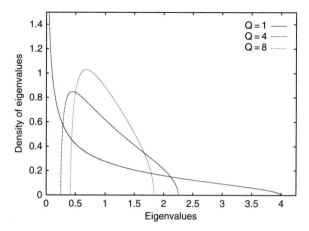

Figure 11.9 Density of eigenvalues of the square random matrix C, as defined in the text (eqn (11.34)) for different values of Q. For $Q > 1$, the function has a cut-off for both small and large values of λ. For $Q = 1$ it diverges as $\lambda \to 0$.

$$\lambda_{min}^{max} = \sigma^2 (1 \pm 1/\sqrt{Q})^2. \tag{11.35}$$

The main characteristics of this density function is that the spectrum has a lower bound and so there are no eigenvalues below $\lambda = \lambda_{min}$ which furthermore tends to zero as $Q \to 1$. The density also vanishes for values of $\lambda > \lambda_{max}$. For finite values of N, these discontinuities become blurred and there is a small probability of finding eigenvalues in the 'forbidden' range.

Figure 11.9 illustrates eqn (11.34) for three different values of Q. This density distribution arises then from the noise associated with the random matrix and, as implied in the introduction, the interesting step is to compare this with the density matrix obtained from analysing real asset price correlation data.

11.5.2 Eigenvalue analysis of stock data

A typical result for an eigenvalue analysis of stock data is shown in Figure 11.10. While the majority of eigenvalues lie within the range predicted by random matrix theory, i.e. $\lambda_{min} < \lambda < \lambda_{max}$, there is a discrete spectrum of larger eigenvalues well outside this range. This is indicative of correlations in the data, which can be explored by looking at the components of the corresponding eigenvectors.

It is found that the eigenvector components associated with the *largest* eigenvalue are distributed more or less uniformly across all the stocks that make up the correlation matrix. The smaller eigenvalues are associated with smaller clusters of company or industrial sectors.

Figure 11.11 shows the eigenvector components corresponding to the *second largest* eigenvalue for a portfolio of 641 stocks traded at the stock exchanges of Paris, London, and New York (the corresponding eigenvalue spectrum is shown in Figure 11.10). The data is grouped into eight different industrial sectors and one can see a clear segregation between (negative) European and (positive) US components. Extending this analysis

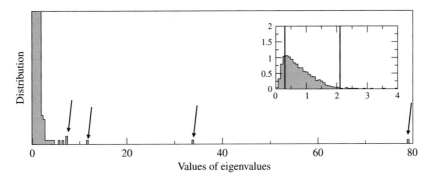

Figure 11.10 Distribution of eigenvalues of the correlation matrix for a portfolio of 641 stocks traded at three different markets: Paris, London, and NYSE (data: daily closure price in US dollars from 30 December 1994 to 1 January 2007). The vertical lines in the inset indicate the region predicted by random matrix theory, $\lambda_{min}=0.2994$ and $\lambda_{max}=2.1107$ (eqn (11.35)). Arrows point to eigenvalues that are outside this region. (Figure reprinted from Coelho *et al.* (2008a), with permission from *Advances in Complex Systems*, World Scientific.)

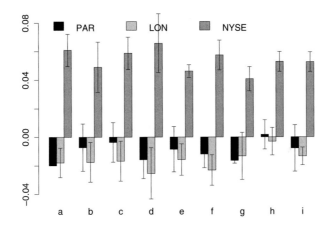

Figure 11.11 The elements of the eigenvectors corresponding to the second largest eigenvalue of Figure 11.10, show a separation of the European data (negative) from the US data (positive). The elements are grouped in markets and nine industrial sectors; (a) basic materials, (b) consumer goods, (c) consumer services, (d) financials, (e) health care, (f) industrials, (g) oil and gas, (h) technology, (i) utilities. (Figure reprinted from Coelho *et al.* (2008a), with permission from *Advances in Complex Systems*, World Scientific.)

to the third eigenvector components, groups together the New York and London data for all eight sectors (Coelho *et al.* 2008a).

This methodology would then appear to be a powerful way of establishing portfolios of assets with specific correlation characteristics, based on the data alone. No other knowledge of a qualitative nature—not even the nature of the company is necessary. More detail of the method can be found in the paper by Plerou *et al.* (2002).

The method has been extended to cover other random matrices arising from Lévy processes. Detail of these developments together with much more detail can be found in the book edited by Burda *et al.* (2005).

11.6   Practical issues

Estimation of volatilities and tail exponents in cumulative probabilities is not always an easy task. The availability of extensive data can be limited, especially in economics. This points to another problem. The approach we have outlined here to assessing risk assumes that any data set is complete/infinite, however, one is usually faced with a limited or finite dataset. So the outcome of any calculation is an approximation to the true value of risk. Indeed the 'true' probability is always going to be different to that computed from the finite dataset. Volatilities and correlations vary over time and any particular sample dataset will not necessarily have the same properties as the complete dataset. So how might an investor estimate the risk in order to guide an investment strategy?

One simple route proposed by Dillow (2012) is to recall the essentially inverse cubic power law for the returns (see Section 6.4). Then if we suppose the annualized volatility is say, 20%, implying that a fall of less than 10% is over the year a certain outcome, then an approximate fit to this data in the tail region is $C_<(x) \sim 1000/x^3$, where x is given in percent. This leads to the outcome that there is a 12.5% chance of a fall of 20% over the year; a 2.7% chance there will be fall of 33% over the year, and a 0.8% chance of a fall of 50%.

Assuming that the volatility scales with the square root of time, or perhaps better estimating the volatility for shorter times, say months, weeks or years, allows an estimate of the risk of similar falls over these shorter times. These risks may seem high and unacceptable to some who will leave money on bank deposit. But that then leaves one clearly open to inflation and currency risk. The point is that such crude estimates are possible, based around the inverse cubic law and, unless one has a pathologic portfolio, they are certainly better estimates of risk than those obtained using Gaussian functions.

11.7   Appendix: Lagrange multipliers

Suppose we seek to maximize[2] a function of two variables, $f(x, y)$, i.e.

$$df = \frac{\partial f}{\partial x} dx + \frac{\partial f}{\partial y} dy = 0.$$

Since dx and dy are independent, we obtain $\frac{\partial f}{\partial x} = 0$ and $\frac{\partial f}{\partial y} = 0$.

Now suppose we want to maximize $f(x, y)$ subject to the *constraint* $g(x, y) = C$, where C is a constant and $g(x, y)$ another function of x and y. In this case dx and dy

---

[2]Note that the requirement $df = 0$ is also the condition for a *minimum* of f, under the constraint g. The method of Lagrange thus gives a *stationary* solution for f under the constraint g.

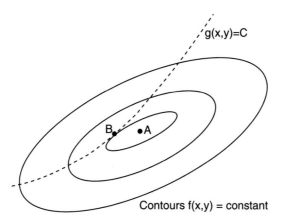

Figure 11.12 The point B corresponds to the maximum of the function $f(x,y)$ under the constraint that this maximum lies on the line defined by $g(x,y) = C$.

are no longer independent, since we now also need to satisfy

$$dg = \frac{\partial g}{\partial x}dx + \frac{\partial g}{\partial y}dy = 0.$$

Therefore we must have

$$\frac{\partial f}{\partial x} / \frac{\partial g}{\partial x} = \frac{\partial f}{\partial y} / \frac{\partial g}{\partial y}.$$

Calling the common ratio λ, we have

$$\frac{\partial f}{\partial x} - \lambda\frac{\partial g}{\partial x} = 0 \text{ and } \frac{\partial f}{\partial y} - \lambda\frac{\partial g}{\partial y} = 0.$$

But these results are exactly what we obtain if we maximise the *Lagrangian* $\mathcal{L} = f - \lambda g$ without further constraints. The solution will now be dependent on the *Lagrange multiplier* λ which is determined using the relation $g(x,y) = C$.

The result has a geometric interpretation, as illustrated in Figure 11.12. Consider the contour map for $f(x,y)$. Without constraint, the solution to our problem is the point A, which is 'at the top of the mountain'. With a constraint, we need to stay on the curve $g(x,y) = C$ and now the solution to our problem is the point B, which is the point at which the line $g(x,y) = C$ tangentially touches the contour of the function $f(x,y)$. (If $g(x,y) = C$ crossed the contour, points on g close to the intersection would result in f being larger/smaller than at the intersection, so f would not be maximized).

For more detail of the method we refer the reader to one of the many mathematical texts, see for example Woolfson and Woolfson (2007).

11.8   Appendix: Wigner's semicircle law

This derivation of Wigner's semicircle law follows the arguments by Bouchaud and Potters (2000). The density of eigenvalues for a square $N \times N$ matrix is given by

$$\rho(\lambda) = \frac{1}{N} \sum_{\alpha=1}^{N} \delta(\lambda - \lambda_\alpha), \tag{11.36}$$

where δ is the familiar Dirac delta function.

Now recall that we seek to compare the empirical results with those which would pertain if the correlation matrix were an $N \times N$ symmetric matrix $\mathbf{H}$, composed of random i.i.d. elements. To compute this density we can use the *resolvent matrix* $\mathbf{R}(\lambda)$:

$$\mathbf{R}(\lambda) = (\lambda \mathbf{I} - \mathbf{H})^{-1}, \tag{11.37}$$

where $\mathbf{I}$ is the identity matrix. The trace of $\mathbf{R}$ is the sum of its eigenvalues:

$$\mathrm{Tr}\, \mathbf{R}(\lambda) = \sum_{\alpha=1}^{N} \frac{1}{\lambda - \lambda_\alpha}, \tag{11.38}$$

where λ_α are the eigenvalues of $\mathbf{H}$.

Using the following identity from complex analysis,

$$\lim_{\varepsilon \to 0} \frac{1}{x - i\varepsilon} = P\frac{1}{x} + i\pi\delta(x), \tag{11.39}$$

where P is the *Cauchy principal value*, we may express the required eigenvalue density in terms of the trace of $\mathbf{R}$ as follows:

$$\rho(\lambda) = \lim_{\varepsilon \to 0} \frac{1}{\pi N} \mathrm{Im}[\mathrm{Tr}\, \mathbf{R}(\lambda - i\varepsilon)]. \tag{11.40}$$

There are a number of ways to proceed. One route, outlined by Bouchaud and Potters (2000), is to set up a recursion relation expressing $\mathbf{R}^{N+1}$ in terms of $\mathbf{R}^{N}$, where $\mathbf{R}^{N+1}$ is obtained by prepending a 0th row and 0th column to $\mathbf{R}^{N}$. From the standard formula for matrix inversion we have,

$$R_{00}^{N+1}(\lambda) = \frac{\mathrm{minor}(\lambda \mathbf{I} - \mathbf{H})_{00}}{\det(\lambda \mathbf{I} - \mathbf{H})}. \tag{11.41}$$

Expanding the determinant in the denominator in minors along row 0, and then each minor into subminors along their first column, gives

$$\frac{1}{R_{00}^{N+1}(\lambda)} = \lambda - H_{00} - \sum_{i,j=1}^{N} H_{0i} H_{0j} R_{ij}^{N}(\lambda). \tag{11.42}$$

Exercise 11.1 The reader might show that eqn (11.42) holds for $N = 2$.

Now let us assume the elements H_{ij} are random variables with zero mean and with variance $\langle H_{ij}^2 \rangle = \sigma^2/N$. This ensures that in the limit $N \to \infty$, the sum remains finite, and that the second term H_{00} can be neglected by comparison with λ. As a result, as $N \to \infty$, we may reduce eqn (11.42) to:

$$\frac{1}{R_{00}^{N+1}(\lambda)} \simeq \lambda - \sum_{i=1}^{N} H_{0i}^2 R_{ii}^N(\lambda) = \lambda - \sum_{i=1}^{N} \frac{\sigma^2}{N} R_{ii}^N(\lambda) = \lambda - \sigma^2 \langle R_{ii}^N(\lambda) \rangle. \tag{11.43}$$

This recursion relation is independent of the precise nature of the statistics that characterize the random i.i.d. elements of the correlation matrix. It is only necessary that the variance σ^2 of the elements is finite. There is a well-defined limit value for the diagonal elements $R^\infty(\lambda)$ given by

$$\frac{1}{R^\infty(\lambda)} = \lambda - \sigma^2 R^\infty(\lambda). \tag{11.44}$$

For $4\sigma^2 \geq \lambda^2$, the solution to this quadratic equation is

$$R^\infty(\lambda) = \frac{1}{2\sigma^2} \left[\lambda \pm i \sqrt{4\sigma^2 - \lambda^2} \right]. \tag{11.45}$$

Using eqn (11.40), noting that $\operatorname{Tr} \mathbf{R} = N R^\infty$, the density of eigenvalues is then

$$\rho(\lambda) = \frac{1}{2\pi\sigma^2} \sqrt{4\sigma^2 - \lambda^2} \text{ if } 4\sigma^2 \geq \lambda^2, \tag{11.46}$$

otherwise the density is zero. This is the 'semicircle' law first derived by the physicist and mathematician Eugene Wigner in the 1950s.

12
Why markets crash

If you see on one hand ice forming on a pool, and on the other a thermometer rising slowly to blood-heat, you may be satisfied in your mind that something is wrong somewhere. Nature cannot make mistakes. Thermometers are equally infallible. Look for the human agency at work on the mercury bulb, for the jet of hot air directed to the instrument. In this parable is explained the market position, and T.B. Smith, who dealt with huge, vague problems like markets and wars and national prosperity, was looking for the hot-air current.

Edgar Wallace, The Nine Bears

In previous chapters we have focused mainly on what a physicist might consider to be systems in a steady state with the addition of market fluctuations. These topics are of keen interest to physicists, but investors are usually concerned with questions of immediate relevance such as:

- Why does a market crash or exhibit speculative bubbles?
- Can I recognize the beginning or end of a crash?
- Can I spot a stock market bubble?
- Can crashes and bubbles be predicted?

These are practical questions that are much harder to address than those we have so far explored. We begin by looking at the work of Sornette and colleagues who have developed a model of crashes, based on ideas originally applied to the study of earthquakes. We then move on to examine insights obtained by Bouchaud and Cont, using ideas rooted in agent models. Both approaches focus on basic concepts and have the benefit of being amenable to analytic treatment.

12.1 Market booms and crashes: some illustrations

Recent years have seen a number of significant booms and crashes in asset markets. Figure 12.1 highlights crashes that have occurred in the London FTSE index since 1986. A number of models have been developed to try to account for this kind of extreme behaviour. The usual approach by economists is to pin a particular reason to a crash, generally some kind of exogenous economic factor. This is not really a very satisfactory approach, since it is never predictive, but rather a post-crash rationalization.

Prior to a crash, when the market is often rising strongly, economists seem to fall into two camps. Some argue that 'it is different this time'. For example, prior to the dot-com bubble 2000 (see Figure 12.1), many economists were saying the new computer technology had allowed changes in efficiency of business operation, so higher

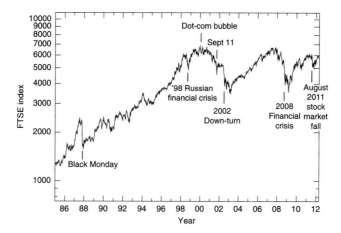

Figure 12.1 Variation of the FTSE index from 1986–2012 on a log-lin scale. A few events that may be identified as crashes have been marked (daily closing price data from Yahoo finance).

economic growth than before could now be sustained. Those who argued otherwise were usually dismissed as being dinosaurs and out-of-touch with reality.

Stock prices are formed as a result of the actions of many traders who are buying and selling particular stocks. The trading process is controlled by a market maker. In days gone by this was a person who matched buyers and sellers. Today, this process is performed computationally in (less than) a few seconds. If the demand for a stock is balanced, such that there are an equal number of buyers and sellers, the price will change little. However, if traders for whatever reason begin to imitate each other and all begin to buy a particular stock, then its price will increase. Continued imitation and buying driven by 'greed' will lead to a continued increase in the price. Such rises can be strong and exist for some time. But at some point, a number of traders may feel that the price has risen beyond what they consider to be a sensible value, become nervous and fearful, and place sell orders. As a result, the price rise will give way to a fall and may as a result crash, and on occasions it does so in quite a dramatic fashion. All practical experienced traders know that 'fear' and 'greed' can drive the action of investors to produce speculative bubbles in this way.

12.2 A mathematical model of rational crashes

In thinking about a mathematical description it should be clear that the time at which a crash occurs cannot be predicted with certainty. Indeed one cannot be sure that a crash will actually happen. A crash can only be characterized in a probabilistic sense by the so-called *hazard rate*, $h(t)$, which is the probability per unit time that the crash will happen in the next instant, given that it has not happened yet. Precisely because it is not possible to deterministically compute the outcome of a crash, some investors will remain invested in an asset whose price is perceived by others to be in a bubble. These investors will expect to be compensated by the higher rate of growth of the bubble and put their faith in the fact that there is a finite probability that the bubble will end without a crash occurring.

Johansen and co-workers, in a series of papers (Johansen *et al.* 1999, 2000) have developed a mathematical theory of stock crashes based around these ideas. We give a short outline of this theory in what follows.

First we note that if we leave aside complicating features such as dividend payments and risk aversion and assume that our investors are purely rational beings, we can invoke what economists refer to as the *Martingale hypothesis*. This essentially states that all other things being equal, the expected asset price $\langle s(t' > t) \rangle$ in the future time t' is the present price $s(t)$,

$$\langle s(t' > t) \rangle = s(t). \tag{12.1}$$

Information pertaining to the future is unknown!

How do we exploit this assertion in circumstances where a crash might be anticipated? Let us now distinguish between the end of a bubble and the time of the crash. A rational investor will assume that the time, t, of the crash has a degree of randomness and this may be characterized by a cumulative distribution function, $Q(t)$, for the time of the crash. The corresponding probability density is $q(t) = dQ/dt$ and the hazard rate is then

$$h(t) = q(t)/(1 - Q(t)). \tag{12.2}$$

The collapse of a bubble is assumed to occur at the most probable time of the crash. The crash itself is assumed to be a so-called jump process, characterized by an integer j whose value is zero prior to the crash and one after the crash.

It follows that

$$\langle dj \rangle_{(t' > t)} = h(t)dt. \tag{12.3}$$

Assuming that at the crash the price drops by a fixed fraction, f, the dynamics of the asset price *before* the crash may be expressed as follows,

$$ds = r_0(t)s(t)dt - fs(t)dj, \tag{12.4}$$

where $r_0(t)$ is a time dependent return.

Forming the expectation value of this equation and invoking eqn (12.3) gives

$$\langle ds \rangle_{(t' > t)} = r_0(t)s(t)dt - fs(t)h(t)dt = 0, \tag{12.5}$$

where the second equation follows from the martingale condition for ds. Hence we deduce that

$$r_0(t) = fh(t). \tag{12.6}$$

Substituting this result back into eqn (12.4) and integrating, we obtain for times that *precede* the crash (when $dj = 0$) the result

$$\ln[s(t)/s(t_0)] = f \int_{t_0}^{t} h(t')dt'. \tag{12.7}$$

Now it is clear, as we would expect, that provided the crash has not happened, then the higher the probability of a crash, i.e. $h(t)$, becomes, the faster the price must increase, as measured by the return r_0, in order to fulfil the rational expectations of our investors. Put another way, in order to be persuaded to hold assets during bubbles, investors require compensation by a return higher than is considered to be normal.

We note here that Sornette in his work also considers a different model, where at the crash, the price falls by a fixed percentage of the price increase above a reference value. This leads to an equation similar to eqn (12.7), except that the log-price is replaced by the price itself. Sornette comments that the model developed in detail above seems to be able to fit bubbles better over longer time periods.

12.3   Continuous and discrete scale invariance

We mentioned above that markets are comprised of traders who are continuously buying and selling stocks. Markets may thus be considered as a network within which individuals and groups interact and affect each others' behaviour. One trader might for example take notice of the action or advice of one or more other traders of this network in order to come to her own decision.

For physicists, such a description is evocative of so-called *spin models* that are used to describe magnetic properties of materials. In the simplest of such models, spins, i.e. elementary bar magnets pointing either up- or downwards, are placed on a square lattice and interact only with their nearest neighbours, but many variations exist.

What is important in our context is that such models can demonstrate what is called *critical behaviour*, where under the right boundary conditions (temperature, external field), all spins begin to align themselves to point in the same direction. A similar situation may arise in an economic or social network, when an initially disordered state may evolve into an ordered one. This ordering can lead to bubbles and crashes.

Near a transition point, a magnetic system has an enhanced sensitivity to both inter-molecular forces and external magnetic fields. Similarly, a social system will, as a transition point is approached, have an enhanced sensitivity to both inter-agent forces via, say gossip, and figures of authority who might give instructions, or spread innuendo. Unforeseen news also can create significant shocks.

In the physical system, critical behaviour corresponds to macroscopic functions, such as magnetisation M having a power-law divergence at a critical temperature T_c of the form $M(T) \sim |T - T_c|^{-z}$, where z is called a critical exponent. The system is said to undergo a *phase transition*.

By analogy it is now assumed that in a social system the hazard rate $h(t)$ can behave similarly, i.e.

$$h(t) = B|t_c - t|^{-b} \tag{12.8}$$

where both prefactor B and exponent b are positive constants. For reasons we shall explain below, b actually must lie between zero and one. Substituting eqn (12.8) into the expression for the price, eqn (12.7) yields for time, t, preceding the crash

$$\ln \frac{s(t)}{s(t_0)} = \frac{-fB}{1-b}(t_c - t)^{1-b} + \frac{-fB}{1-b}(t_c - t_0)^{1-b} = \frac{-fB}{1-b}(t_c - t)^{1-b} + \ln \frac{s(t_c)}{s(t_0)}. \tag{12.9}$$

Here we have introduced $s(t_c)$ as the price at the critical time $t = t_c$, conditional on the crash not having happened yet. Note that the price $s(t_c)$ must be finite since if it were not so, then the theory would unravel because rational agents could anticipate the crash. This implies that $1 > b > 0$.

From eqn (12.8) it follows that

$$\frac{d \ln h(t)}{d \ln |t_c - t|} = -b. \tag{12.10}$$

This states that the hazard rate is *self-similar* close to the critical time, t_c, with respect to the dilation of the distance $t_c - t$; the relative variations of $\ln h(t)$ with respect to $t_c - t$ are independent of the time t. The reader can easily see that a similar relationship exists between $\ln \frac{s(t)}{s(t_c)}$ and $t_c - t$ since the exponent $1 - b$ is also constant.

Considering our market, we can imagine that as the critical point is approached, larger and larger groups are developed who are 'bearish', that is consist of traders who 'feel' the price is going to fall imminently. The numbers of bearish traders eventually outweighs the bullish traders who remain convinced a further rise in the price is possible. At the point when the last bullish trader switches to bearish, the actions of all traders become correlated and the price collapses.

For physicists this is analogous to a phase transition, as mentioned earlier. The physicist Kenneth G. Wilson was awarded a Nobel prize for his work that showed how power law solutions of the type derived above emerged from renormalization group equations or continuous scale invariance (Wilson, 1979). This implies that a function of a thermodynamic system, say the magnetism $M(x)$, reproduces itself up to a constant when the scale x is changed to say κx. Thus

$$M(x) = \mu M(\kappa x). \tag{12.11}$$

This equation is solved by the power law relation

$$M(x) = Cx^{-\alpha}, \tag{12.12}$$

together with the condition

$$\kappa^\alpha \mu = 1, \quad \text{i.e. } \alpha = -\frac{\ln \mu}{\ln \kappa}. \tag{12.13}$$

In 1996 Feigenbaum and Freund, followed later by Johansen *et al.* (1999), pointed out that in many systems there is no reason for the exponent α to be real. For example, materials with a hierarchical structure exhibit a *discrete scale invariance* which needs to be taken into account when describing their failure mechanisms.

This follows when we write eqn (12.13) as follows,

$$\lambda^\alpha \mu = 1 = \exp(2\pi i n), \quad \text{thus } \alpha = \alpha_n = -\frac{\ln \mu}{\ln \lambda} + \frac{2\pi i n}{\ln \lambda}, \tag{12.14}$$

where n is a integer.

Continuous scale invariance is equivalent to the case where $n = 0$ and the solutions with $x \neq 0$ are discarded. However noting that $\Re[x^{-z+i\omega}] = x^{-z} \cos(\omega \ln x)$ where $\Re$

denotes real part, we see that if non-zero values of n are allowed, a series of log-periodic oscillations are superimposed on the power law, viz

$$(t_c - t)^\alpha = (t_c - t)^{\alpha_0} \left(1 + \sum_{n=1}^{\infty} c_n \cos \left[\frac{2\pi n}{\ln \lambda} \ln |t_c - t| \right] \right), \tag{12.15}$$

where the c_n are Fourier coefficients, to be determined by constraining boundary conditions, and $\alpha_0 = -\frac{\ln \mu}{\ln \kappa}$.

Exercise 12.1 The reader may wish to show the above scaling explicitly.

Oscillations of this nature have been observed in analysis of both financial and earthquake data. For the latter they were found in measurements of the concentration of Cl^- ions in ground water close to the epicentre of an earthquake, where the concentration changes reflect strain field changes in the ground prior to a quake (see for example the discussion in Voit, 2001).

Fits of stock market data were performed by both Feigenbaum and Freund (1996) and Sornette and Johansen (1997) using the approximate solution, including only one Fourier coefficient C,

$$h(t) \approx B|t_c - t|^{-b} + C(t_c - t)^{-b} \cos[\omega_1 \ln(t_c - t) + \phi] \tag{12.16}$$

for the hazard rate. For a comprehensive account of this approach see the book by Sornette, 2003. This results in the following expression for the stock price as a function of time

$$s(t) \simeq A_1 + B_1(t_c - t)^{1-b} + C_1(t_c - t)^{1-b} \cos[\omega_1 \ln(t_c - t) + \phi_1]. \tag{12.17}$$

Here $A_1, B_1, C_1, \omega_1, b$ and ϕ_1 are fit parameters and the scale factor λ is obtained from the frequency ω_1 of the oscillation,

$$\omega_1 = \frac{2\pi}{\ln \lambda} \text{ i.e. } \lambda = \exp[\frac{2\pi}{\omega_1}]. \tag{12.18}$$

Based on these ideas, a number of groups have fitted different crashes from the Dow Jones, S&P, and the Nikkei indices. Some good fits have been found, but in other cases the predicted crash did not occur on schedule. It may occur before the predicted moment; it may not occur at all. Since the theory is based on probabilistic methods perhaps one should not be too surprised at this. A fit to S&P 500 data prior to the October 1987 crash, as originally analysed by Sornette *et al.*, is shown in Figure 12.2.

However, it is often said that with so many parameters one can 'fit an elephant' to the theory and some researchers have suggested that the fits that have been found are not statistically significant. Despite all these reservations, evidence for the validity of the theory continues to be accumulated. It is probably the only theory that can claim a predictive character at this time.

One additional point noted by Sornette is that these periodic patterns are not dissimilar to certain patterns followed empirically by technical traders or so-called

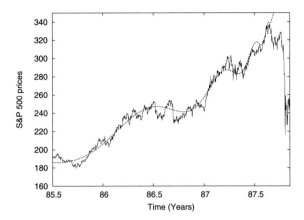

Figure 12.2 The S&P 500 data prior to the stock market crash in October 1987 may be described by the log-period oscillations of eqn (12.17).

chartists. Many of these patterns are simply empirical in origin. A few are characterized by mathematical functions, including Fibonacci numbers and various wave like patterns. One such pattern is based on the logarithmic spiral. In polar coordinates, its radius r is given by $r = r_0 \exp(a\theta)$ as a function of the polar angle θ and a growth-rate $a > 0$. Since $x = r \cos \theta$, it follows that this function crosses the x axis when $\theta = n\pi$. Alternatively, re-expressing these values in terms of the set of intersections, $\{x_n\}$, on the x axis one obtains $x_n = (-1)^n \exp(an\pi)$. This is seen to be a discrete geometric series similar to that expressed by eqn (12.18), i.e. an oscillation together with an exponential change in scale.

However, as Sornette himself notes, there is a difference between a theory based on tools of statistical physics and mere pattern recognition. Nevertheless, the results derived here will, no doubt, offer some support and confidence to those who rely on chartist tools for their investment strategies. The methods outlined here have been applied by Zhou and Sornette (2006) to study house price crashes, a topic we shall turn to in Section 13.2

12.4   Agent models

Another approach to the examination of stock market crashes has been provided by Bouchaud and Cont (1998). In a very nice paper that is well-worth reading, they examined the properties of a simple agent model that was amenable to analytic investigation. They noted that many different types of markets exhibited similar characteristics of the type already discussed, i.e. non-Gaussian statistics over short timescales, time dependent volatility autocorrelations with slowly decreasing power law tails and over the long time, a log price that increases on average linearly with time. Recognizing these common features, together with the fact that it is impossible to get access to the detailed behaviour of every single agent, Bouchaud and Cont proposed a phenomenological model where trading was characterized simply by two types of agents, namely chartists or noise traders and fundamental traders/investors (see Sections 2.2.2 and 2.2.1). All the demand for a stock was assumed to arise from these two groups of traders. In this way the authors hoped to capture the basic individual and collective behaviour of the agents in the manner of a mean field theory used to describe cooperative phenomena in statistical physics. We give below an outline of their approach (for details see Bouchaud and Potters (2000)).

At time t, the asset price s will be a function of the excess demand, $\phi = \phi_+ - \phi_-$ for the asset. Demand is positive when the amount of stock wanted by buyers, ϕ_+, exceeds that being sold, ϕ_-, and vice versa; it is negative when the stock being sold exceeds that which is being bought. Bouchaud and Cont considered an approximately linear variation of S with excess demand, thus

$$\frac{ds}{dt} = u(t) = \frac{\phi}{\lambda_d}, \tag{12.19}$$

where we have introduced the instantaneous price-change $u(t) = \frac{ds}{dt}$. The constant λ_d may be interpreted as a measure of the *market depth*, i.e. the excess demand per price change. If λ_d is large, the market is able to make only small price changes as demand changes. On the other hand, small values of λ_d may lead to large fluctuations in price.

It is usually the case that deals or trades are conducted by a market maker. This may be a person at the end of the telephone or, more usually nowadays, a computer. The function of the market maker is to match buyers with sellers and, in this way, absorb the demand (or supply). If we assume the market maker behaves in a symmetric manner towards buying and selling activity, the effect is simply modelled, viz

$$\left.\frac{d\phi_\pm}{dt}\right|_{\text{market makers}} = -\gamma\phi_\pm, \tag{12.20}$$

where $1/\gamma$ is the characteristic time for deals to be completed. In the mid-nineteenth century, such times were typically of the order of a number of days. Today, with modern computers, these times are of the order of a few minutes, or even less.

How do the traders contribute to the demand? Fundamental traders have in their mind an estimate of what they consider to be fair value, s_0. If the price rises above this they anticipate the demand will fall; if the price is lower than the fair value, they anticipate the price will rise. On the other hand, noise or technical traders seek to anticipate future trends based on past performance. A rising trend may be expected to continue and so add to demand for an asset; similarly a falling trend may reduce demand. The contribution to the demand from such a trend will be proportional to something akin to

$$\hat{u} = \int_{-\infty}^{t} K(t - t')u(t')dt', \tag{12.21}$$

where the kernel K represents how the agent weighs the price changes at different times. If we assume the agent response time is short, then to leading order, K may be replaced by a δ-function and the contribution of the term to the demand is made by simply replacing $\hat{u}$ by its instantaneous value $u(t)$. The total rate of change of demand from traders then reduces to

$$\left.\frac{d\phi}{dt}\right|_{\text{traders}} = \alpha(u)u(t) - k(s - s_0). \tag{12.22}$$

The first term on the RHS is due to noise traders, and the second due to fundamental traders, where the constant k is a measure of the force that leads the price S to revert to the fundamental value S_0.

Recognizing a little psychology, which is that to lose money is worse than gaining less than one might have envisaged, one can expect the pre-factor α to depend on the instantaneous price-change u, and the simplest assumption consistent with this empirical observation is to let $\alpha \to \alpha - \alpha' u$, where α and α' are now both positive constants. This particular form of α mimicks that at some point traders begin to 'jump off the bandwagon' and sell, so the momentum begins to decrease.

Agents also look to past variations of volatility to gauge risk, leading to a second contribution to demand. Assuming all agents behave in a similarly risk-averse way, then the contribution from this term will be a decreasing function of the volatility. If we again assume that the memory of the agents is short, then we may use the instantaneous volatility, which is proportional to u^2 (with proportionality constant β), in estimating this contribution to demand. Thus:

$$\left.\frac{d\phi}{dt}\right|_{\text{risk}} = m - \beta u^2, \tag{12.23}$$

where m is the change in ϕ at $u = 0$.

The numbers of agents will also change in response to new information, a need for cash, or particular investment strategies. This is captured through a random contribution to the changing demand,

$$\left.\frac{d\phi}{dt}\right|_{\text{random}} = \eta(t). \tag{12.24}$$

The random term, $\eta(t)$ is normalized as follows,

$$\langle \eta(t)\eta(t') \rangle = 2\lambda_d^2 D\delta(t - t'). \tag{12.25}$$

This normalization ensures that D measures the susceptibility of the market return, u, to external shocks, such as the arrival of news or a sudden increase in the numbers of traders. Indeed, through this dependence arise the fat tails in the probability distribution for returns that we have described earlier (see also Figures 6.3, 9.3 and 9.4). However, for simplicity, here we ignore this added complication and take D to be a constant. We will return to this issue in Section 12.4.4.

Bouchaud and Cont also note that there will be other contributions to demand from option traders, since as we have learned in Chapter 8, the purchase of options used to hedge or insure positions gives rise to additional trading of the underlying asset. Again this aspect is ignored.

So adding together all the contributions, eqns (12.20), (12.22), (12.23), (12.24) and using equation (12.19) yields:

$$\lambda \frac{d^2 S}{dt^2} = \frac{d\phi}{dt} = -\lambda \gamma u + \alpha u - \alpha' u^2 + m - \beta u^2 - k(S - S_0) + \eta(t). \tag{12.26}$$

Note that all the coefficients on the RHS are positive. It is now interesting to examine some different limiting situations where this equation can be analysed analytically.

12.4.1 Liquid and illiquid markets

Consider first the risk neutral case where $\alpha'=\beta=0$. Eqn (12.26) now reduces to a linear second order differential equation characteristic of a damped harmonic oscillator, driven by a random force $\eta(t)/\lambda$,

$$\frac{d^2 S}{dt^2} + \psi \frac{dS}{dt} + \frac{k}{\lambda}(S - S_1) = \frac{\eta(t)}{\lambda}, \tag{12.27}$$

where $s_1 = s_0 + m/k$ and the 'damping constant' ψ is given by $\psi = \gamma - \alpha/\lambda_d$. For *liquid* markets where γ and λ_d are large (corresponding to short characteristic times and large market depth) and $\gamma > \alpha/\lambda_d$, the solution and hence this market model is stable. The price oscillates around the price S_1, which is greater than the fundamental price, S_0.

Exercise 12.2 The reader may show that the time correlation function consists of two exponentials with correlation times

$$\tau_1 \simeq 1/\psi \text{ and } \tau_2 \simeq \tau_1/\varepsilon \text{ where } \varepsilon = \frac{k}{\psi^2 \lambda} \tag{12.28}$$

and amplitudes A_1 and $A_2 \simeq \varepsilon^2 A_1$.

It follows that on a timescale τ_1 the amplitude of the correlation function falls to a very small value and $\tau_1 \sim 1/\psi$ may be equated to the correlation time of liquid markets, which is of the order of a few minutes, as noted above. On timescales $\tau_1 \ll t \ll \tau_2$ the stock price behaves as a random walk with volatility $2D\tau_1^2 = 2D/\psi^2$, before feeling the pull of the harmonic force back to the 'fundamental' price s_1.

There will usually be some disagreement about the fundamental price. When the market is quiet (unlike the state at the time of writing this!) traders probably agree about the fundamental price to within 10-15% and a typical variation of stock price during a year may be of this order of magnitude. So we might assume that the timescale τ_2 is of the order of a year. Hence from eqn (12.28) we deduce that $\varepsilon \sim 10^{-4}$. So for short timescales—less than a year—the harmonic force ($k = \epsilon \psi^2 \lambda_d$) that seeks to confine the stock price can be ignored.

For illiquid markets where $\psi < 0$, the solutions are unstable, leading in the model to solutions that increase or decay exponentially. These may be interpreted as speculative bubbles. But now it is no longer valid to neglect the higher order terms that characterize risk aversion.

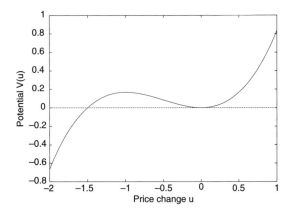

Figure 12.3 The potential $V(u)$ features a local maximum at $u^* = -\psi\lambda/b$ and a minimum at $u = 0$ (for the graph we have set $\psi = \lambda = b = 1$).

12.4.2 Risk aversion induced crashes

Let us now explore the effect of the non-linear terms previously neglected. If we limit our timescale to less than a year, we may, following the above analysis, neglect the harmonic force. Now eqn (12.26) reduces to

$$\frac{du}{dt} = \frac{m}{\lambda} - \psi u - \frac{b}{\lambda}u^2 + \frac{\eta(t)}{\lambda}, \tag{12.29}$$

where $b = \alpha' + \beta$.

This is a classic Langevin equation (see Section 5.1) from which it may now be seen that the instantaneous price change, u, fluctuates within the confines of a potential $V(u)$, where

$$\frac{\partial V}{\partial u} = -\left[\frac{m}{\lambda} - \psi u - \frac{b}{\lambda}u^2\right]. \tag{12.30}$$

At this point one needs to set the constant term, m/λ. This term represents an average trend and leads to a biased random walk, but the actual value of m (provided it is larger than some finite negative threshold) makes no material difference to the qualitative picture developed below.

So setting m to zero, as done by Bouchaud and Cont, the potential V is given by

$$V(u) = \frac{\psi}{2}u^2 + \frac{b}{3\lambda}u^3. \tag{12.31}$$

If the constants ψ and b are both positive, this function has a local minimum when $u = 0$ and a local maximum at $u = u^* = -\psi\lambda/b$, as shown in Figure 12.3.

Clearly if $u < u^*$, then the motion becomes unstable and u can descend to large negative values, leading to a large drop in price s. So we see that an asset may have a price that fluctuates around some initial value in the vicinity of the potential well (the 'normal' region), but after a series of random jumps the instantaneous return may reach the local maximum, cross into the 'unstable' region and collapse to a large negative value in a finite time.

The model says nothing about how the price recovers from this price crash. But of course from our earlier discussion we can expect that fundamental investors come in and buy assets with a price below the fundamental value, giving rise to a recovery in the price.

Note that this model does not produce any precursors ahead of the crash, for example in the manner of the log-periodic oscillations discussed in Section 12.3. A crash occurs simply after a series of random jumps that together lead the motion to enter the unstable region, $u < u^*$.

It is possible to compute the price as a function of time as this region is entered. In this case, the noise term on the RHS of eqn (12.29) can be neglected.

Exercise 12.3 Continuing also to neglect m, the reader should show by inspection that

$$s(t) = s^* + \frac{\lambda_d}{b} \ln \left[\exp \psi (t_f - t) - 1 \right], \tag{12.32}$$

where the time t_f is the time of the crash at which point the price, $s(t)$, diverges to $-\infty$. s^* corresponds to u^*.

Bouchaud and Cont fit this function to a crash of the S&P index that happened in October 1987. Whilst a reasonable fit could be obtained, the value deduced for ψ proved to be an order of magnitude smaller than one would expect, given τ_1 should be of the order of a few minutes or so.

To circumvent this difficulty, Bouchaud and Cont noted that the discussion so far assumed that the market participants, i.e. the noise traders and fundamental agents have essentially no memory (see the discussion following eqn (12.21)). They argued that in reality, whilst the market maker may react with a short response time, the traders take account of market moves over a longer timescale. Assuming that the kernel in eqn (12.21) is exponential, with a decay time $1/\Gamma$ that is much greater than the characteristic time associated with the market maker $1/\gamma$, Bouchaud and Cont find that the form of eqn (12.7) remains unchanged, except for the coefficients,

$$S(t) = S^* + \frac{\lambda_d \gamma}{b \Gamma} \ln[\exp(\Gamma(t_f - t)) - 1]. \tag{12.33}$$

So the scale over which the crash develops is now determined by Γ and Bouchaud and Cont find, by fitting the various parameters, that $1/\Gamma \sim 220$ minutes or approximately half a day. This is consistent with the earlier assumption that traders take account of market moves over time scales longer than the few minutes/instantaneous scale associated with market makers.

12.4.3 Speculative bubbles

Does this model reveal any insight into the formation of speculative bubbles? Let us suppose that the noise traders are very active, so that α is large and positive; alternatively, or in addition, assume that the liquidity factor λ is small. In this case it is possible for the coefficient $\psi = \gamma - \alpha/\lambda_d$ to be negative. The nature of the solution to eqn (12.29) thus changes. The potential $V(u)$ now has a local maximum when

$u = 0$ and the local minimum $u^* = -\psi\lambda_d/b$ is now positive. In this case, the price increment moves around the position u^* and over a time t the price increases roughly by an amount u^*t. So a price bubble forms. This cannot increase forever; eventually the fundamental traders will enter the market and sell, causing the bubble to collapse after a time $t_b \propto \tau_2$, where the parameter λ_d is now replaced by Γ (see eqn (12.28)). The lifetime of the bubble is then $t_b \sim \psi\Gamma/k$ and the stronger the constant, k, the shorter the lifetime of the bubble. One might say that k is a measure of the number of fundamentalists trading in the market.

12.4.4   Anomalous diffusion

One of the more intriguing aspects of this theory is not just that it has two regimes, a Gaussian or normal regime and a 'crash' regime, but that the model suggests that the dynamics of the crash are determined by parameters that describe the normal regime. Bouchaud and Cont therefore assert that non-trivial correlations may be observed within financial data which are not simply random. But do any of these correlations give rise to the log-periodic oscillations proposed by Sornette and discussed in Section 12.3? As formulated above, the theory does not seem to do so. However, by introducing non-Gaussian noise, of the kind discussed in Chapter 10, where the diffusion coefficient is now dependent on the price return, a tentative link may be made.

The argument goes as follows. Suppose the price return reaches a value close but above u^* at which a crash would happen (remember that u^* is negative). Assuming a constant diffusion coefficient, then the time, t^*, for a crash to occur may be computed using the Arrhenius law,

$$t^* \propto \exp(V^*/D), \tag{12.34}$$

where V^* is the height of the potential function at $u = u^*$.

Now suppose that the susceptibility of investors to news is heightened in this pre-crash regime and the effective diffusion coefficient D in eqn (12.25) increases to a new and higher value $D + \delta D$ where $\delta D \sim u^2$. If D increases in this way at each such large fluctuation at time t_n, then the time between such events is decreased by a factor

$$(t_{n+1} - t_n) \sim (t_n - t_{n-1})\exp(-\frac{V^*\delta D}{D^2}). \tag{12.35}$$

The time difference between two events is thus seen to follow a geometric series which leads approximately to log-periodic behaviour (see the discussion at the end of Section 12.3). However, in this approach, the log-periodic behaviour arises when $u > u^*$ and not as the result of the accumulation of numerous events of various magnitudes, as in the theory of Sornette.

12.5   What happens after a crash?

The theories described above represent important and interesting steps towards understanding of asset price dynamics. However, despite evidence for some universality, there remain considerable difficulties in predicting the future behaviour or even detecting the onset of price crashes. So there is more to be uncovered within these complex systems.

An important complementary question that might be asked by any investor is: are there any indicators of the end of a crash? How might one judge that one is emerging out of a crash regime and into a more 'normal' market regime? Bertrand Roehner (2000, 2001) noted that a trend reversal after a crash will correspond to a change in trader attitude from bearish to bullish behaviour and fear of the future is replaced by optimism. Roehner's idea is to seek out an indicator that essentially is a measure of this uncertainty about future prospects.

He proposes interest rate spreads for a sample of bonds as a proxy for this uncertainty, assuming it to be a statistical measure of lenders' opinions of the future. From a study of eight major stock market crashes in the US between 1857 and 1987, he suggests that markets begin to recover when interest rate spreads begin to decrease.

At the time of writing, many investors are concerned about the spread of interest rates across the member states of the eurozone, following the financial crash of 2008 and the rescuing from bankruptcy of banks across Europe by European governments. The financial markets are presently very volatile and consumer confidence in the markets is not high. It remains to be seen if we shall have another financial crash before the insolvency of states such as Greece is resolved. More work is needed in this area.

A number of papers have been published that purport to simulate a financial market. Whilst these usually reproduce the statistical properties of a market, as discussed in Chapters 2 and 6, extracting insight into the dynamics is not so easy, and much more in the way of empirical analysis and understanding is needed before detailed computer models will be reliable. In this sense, the problem is similar to that of climate change. Computer models can make predictions but it is important to know that the models are rooted in reality and capture the empirical phenomena correctly.

13
Two non-financial markets

Someone once asked me why women don't gamble as much as men do, and I gave the common-sensical reply that we don't have as much money. That was a true but incomplete answer. In fact, women's total instinct for gambling is satisfied by marriage.

Gloria Steinem

Are stock traders really only well-paid punters? Should house buyers not only compare prices across a city, but across the globe to spot potential bubbles? In this chapter we present recent findings concerning online betting markets and house markets.

13.1 An analysis of online betting markets

Traditionally, if one wanted to bet on a certain outcome of an event, such as a horse-race or a soccer match, the bet would be placed with a bookmaker. The bookmaker sets the *odds* deemed appropriate for the expected probability of the outcome and the customer can place bets against those odds. If a bet is offered, or *laid*, with odds for example at 3 to 1, the customer who *backs* the bet, can expect to win three times the money he or she placed on the bet with a probability of 1/4. The bookmaker will consequently keep the placed money with a probability of 3/4. Bookmakers can expect to make profits in the long run by providing odds to their customers that slightly overestimate the true probabilities of that given outcome.

With the ever increasing numbers of facilities that the World Wide Web offers, it is no surprise that betting may also now be carried out *online*. In this case, the role of a single bookmaker is abolished and the online betting exchange operates in a manner more akin to a stock exchange. *Users* are now able to play both the roles of punter (customer) and bookmaker by choosing to either back an outcome or lay a bet that can be backed by another user. They have access to a double auction order book which is similar to one used in a regular financial market, except that the *bid* and *ask* columns are now called *back* and *lay* columns. An important difference to financial markets is, however, that a betting market has a definite conclusion. The odds will inevitably move towards infinity or zero as the outcome of the event on which betting took place becomes a certainty, e.g. as the final whistle blows in a soccer match. This is a feature also found in so-called *prediction markets*, i.e. speculative markets where assets traded have a value that is tied to a particular event (i.e. will the next US president be a republican?). Some research suggests that a study of such markets provides excellent predictions of the event outcome (see, e.g. Berg and Rietz, 2006).

We shall, in the following, present some recent results from an analysis of 146 soccer matches that took place as part of the 2007/2008 Champions League tournament (Hardiman *et al.* 2010, Hardiman *et al.* 2011). The data was obtained from the online

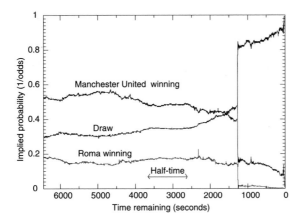

Figure 13.1 Evolution of the implied probability (related to the reciprocal of match odds) for the three possible outcomes, i.e. Manchester United winning, Roma winning, or a draw. As this Champions League 2008 match draws towards its conclusion, the implied probabilities for the outcomes tend towards zero and one, respectively. We see a large change in probabilities for each outcome at the t=1400s mark, corresponding to a goal scored by a Manchester United player, Carlos Tévez, which lead to the final score Manchester United 1 - Roma 0. Since Betfair (http://www.betfair.com) provides data at one second intervals, the displayed data is made up of about 6500 data points.

betting company Betfair, which according to its homepage records 'five million transactions per day, more than on all European stock exchanges combined'. As in financial markets, the distributions of returns are fat tailed (*cf.* Section 6.3) and volatility is a long memory process (*cf.* Section 2.1.3). However, since betting takes place also at the half-time interval where there is no game activity, we also obtain information on the statistics of purely trader-driven odds movement.

Betfair data for backing and laying bets is available at one-second intervals and includes the value of the odds at which the last bet was matched during that second. The odds on Betfair are not given in the traditional gambling parlance of say '2-to-1' or '3-to-1' but as 3.0 and 4.0 instead. From this one can compute 'implied probabilities' $p(t) = 1/\text{odds}(t)$ and use these to compute log-returns, defined as usual as $r(t, \delta t) = \ln \frac{p(t+\delta t)}{p(t)}$, *cf.* eqn (2.6).

Figure 13.1 shows the variation of these implied probabilities over the course of the soccer match Manchester United vs. Roma that took place on 9 April 2008. Fluctuations of odds are visible throughout the entire match, although at a much reduced level during the half-time break (note however that trading volume, which is also recorded on the Betfair website, remains significant also during half-time). The goal scored by Manchester United towards the end of the match may be clearly identified by large changes in the probabilities associated to the three possible match outcomes of Manchester United winning, Roma winning, or a draw.

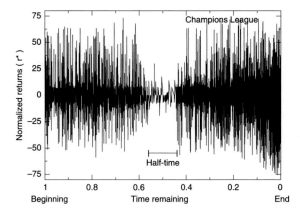

Figure 13.2 The (normalized) log-returns grow larger in magnitude as the match draws to a conclusion and are greatly reduced during the half-time interval. The data shown is an average over a set of 146 soccer matches from the 2008 Champions League football tournament, where the returns in each match have been normalized by the standard deviation of returns in that match.

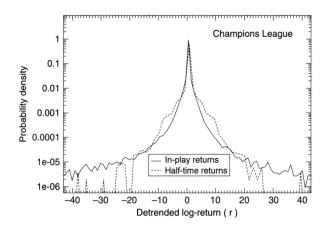

Figure 13.3 The distribution of (detrended) log-returns for the 2008 Champions League data shows that the in-play data exhibits fatter tails than half-time interval data. While the latter is due to trader noise, in-play data reflects the activity on the pitch. In order to compute this distribution, the normalized return data of Figure 13.2 was divided by the local volatility average. This *detrending* procedure takes into account the fact that volatility is not constant for the duration of a match, but increases with time remaining to the end of the match.

Similar to the study of say a portfolio of FTSE 100 shares, one may wish to look at a combined set of soccer matches. As in the treatment of financial data, this requires that the data for each match is normalized by the respective standard deviation (cf. Section 11.3). Figure 13.2 shows such normalized returns for the combined data of 146 soccer matches from the 2007/2008 Champions League football tournament. The character of the half-time data is clearly different to the one of the in-play data. While the in-play data reflects the changing probabilities for the outcome of a match due to the game-play and scoring in the match, the half-time data is representative of endogenous market behaviour, i.e. trader interaction in the absence of external news.

The probability densities of the returns of both in-play and interval soccer data feature fat tails, as shown in Figure 13.3, but this is more pronounced for the in-play data with the possibility of sudden events (goal scores). Similarly, the character of different sport disciplines may result in distinctly different distributions of returns. Scoring a point in a tennis match is generally not having such a dramatic influence on the probability of winning the match than it is for the case of a goal in soccer. Consequently, the tails in the distribution of returns for soccer are distinctly fatter than for tennis (Hardiman *et al.*, 2011).

As in financial data, long-range correlations exist in the magnitude of returns (volatility), with a Hurst exponent of about 0.6 for both in-play and interval data. Since in the interval little news is hitting the betting market, one may conclude that the long-range correlations are due to the trader-driven noise. Trading volume in the betting market is, however, unlike what has been observed in financial markets (e.g. Lobato and Velasco, 2000), not a long memory process, i.e $H \simeq 0.5$.

The rising popularity of online betting with its increasing economic impact is undoubtedly a great incentive for further detailed studies of its underlying mechanism.

13.2 House price dynamics

People everywhere need a place to live and provide shelter for their families. Whether they are able to do this in much of the world depends on their ability to either buy or rent a house or apartment. Since rented accommodation is owned by a landlord, the provision of affordable rented accommodation also depends ultimately on the price of housing. As a result, house prices are of great interest and concern to many people. However, how do people judge whether prices in the market place are good value or not?

During early 2006, the press headlines in the UK were all saying that prices will continue to rise across the UK and in London in particular (cf. Figure 13.4). In October that year, the UK Royal Institution of Chartered Surveyors reported that prices were rising at their fastest rate for four years. The Financial Times of 4 November 2006 contained a feature suggesting that social forces, such as immigration, were changing the very fabric of the house market. That seemed to echo the view of commentators who at the peak of the dot-com boom were suggesting that the nature of the economy had changed and price to earnings ratios (P/E ratios, see Section 2.2.1) that had reached very high values were now the norm. Simultaneously, the UK Abbey National Building Society launched a new mortgage scheme, enabling homebuyers 'who cannot save up [the deposit] quickly enough' to borrow up to five times their joint salary. Elsewhere in the same newspaper could be found an article indicating that the use by sellers of the 'sealed bid', a device used to extract the maximum price, was making a comeback. Of this, one commentator remarked 'It's a piece of crass opportunism with "top of the market" written all over it'. Few economists expressed caution and by and large buyers continued to push the market up.

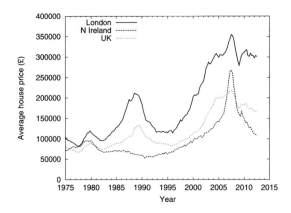

Figure 13.4 Evolution of average house prices (in £ sterling) in the United Kingdom over the period 1975–2012. Shown in addition are the house prices of London and Northern Ireland. (Data obtained from Nationwide Building Society).

In Dublin on 19 September 2007, Ireland's largest property advisory firm, the Sherry FitzGerald Group, released its 'News for the Irish property market':

'[The past 20 years have] been an extraordinary period. Each year some new reliable source has predicted the demise of this market, gained the headlines only to be quietly proven wrong by the zealous growth rates. This trend continued right up to present day. Despite 10 years of phenomenal growth and 6 consecutive interest rate increases, 2006 was one of the most remarkable years for the Irish property market. The average price of a second-hand property in Ireland rose by 18% during the 12-month period. A record high of 93,400 residential units have been completed in 2006. Economic growth is projected to be a very healthy 5.4% this year and approximately 4% next year. Indigenous home-buyers will also be enhanced by a vast number of migrants who are choosing to come and live in Ireland. Last year alone some 87,000 immigrants came to live in Ireland, fueling demand for almost 30,000 residential units. In short, we have a young and growing population in a strong economy with more people at work than ever before. How much longer can this last? The answer is that all indicators point to it being sustained for a very long time indeed. Current estimates suggest price inflation will slow to single digit figures in 2007 and into 2008.'

This statement was released just a few months before the sudden market downturn. In fact in 2007 London prices fell by 5%. Dublin property prices similarly fell by 7%, yet GDP growth remained around 5% in 2007 but fell to 0.7% in 2008 and to -7% in 2009.

Why did the assessment turn out to be so wrong? For a very simple reason: everyone forgot the role of speculation as the main market engine in such episodes. As prices go up, the percentage of purchases made by investors (that is to say for the purpose of reselling fairly quickly with a profit) increases steadily and the role of the so-called fundamentals that control the supply-demand balance in normal times becomes more and more negligible and irrelevant. A self-sustained but highly unstable system results, and as soon as the profit incentive disappears, the market is set to tumble. It is a typical *collective delusion effect*. It is based on the recurrent effects of such collective delusions that an analysis of housing markets is possible.

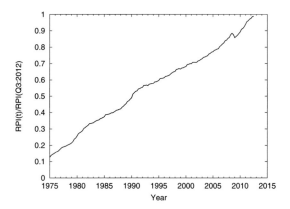

Figure 13.5 Variation of the UK retail price index (RPI) from 1975 to 2012, where the index is computed such that it equals unity at the third quarter of 2012. (Data obtained from Nationwide Building Society).

Data for house prices in the UK and Northern Ireland can be obtained from the Nationwide Building society website[1], and Figure 13.4 is based on data over the period 1975 to 2012. Peaks that occurred in the UK during the late 1980s and 2007 are clearly visible.

The peaks are brought out more clearly if the actual prices are renormalized using the retail price index (RPI) to obtain the 'real' house price, which equals the actual house price divided by the RPI. The RPI, shown in Figure 13.5, is a measure of inflation, obtained from the change in the cost of a defined basket of retail goods and services, compared to a reference date. Normalization of the data shown in Figure 13.4, using the third quarter of 2012 as the base for the RPI results in the so-called 'real prices', as shown in Figure 13.6. A log-linear plot brings out the long-term linear trend.

To a first approximation we may say that the shape of the speculative peaks in these examples is roughly similar and such similarity has also been observed in house price peaks elsewhere, including Dublin (Richmond 2007, 2009), the West of the USA, Australia (Roehner 2006), and Paris (Roehner 1999). Recognizing this similarity, both authors independently published predictions for the evolution of house prices in the west of the USA, London (UK), and Dublin (Ireland). The predictions were derived from the clear pattern set by speculative price peaks that had occurred in former decades.

In a joint publication, Richmond and Roehner (2012) reassessed their predictions by comparing them with the data describing the price evolution over the previous 5–6 years. In both cases the differences between projections and observed prices (as of end of 2011) was less than 15%, (14% for the average of the London-Dublin case and 8.5% for the west of the USA).

---

[1] http://www.nationwide.co.uk/hpi/historical.htm

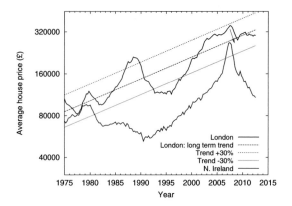

Figure 13.6 Normalizing the house prices shown in Figure 13.4 by the retail price index (third quarter 2012 prices) and plotting the data on a log-lin scale reveals the similar nature of the speculative peaks. Trend lines are shown for the London data. The arrow indicates the prediction made in 2006 of a large price drop.

The more closely any pattern repeats itself in the past the more likely it is that a new episode will follow the same pattern. It is easy to predict, after seeing the sunrise at about 5:30 for a few days, that it will occur at about the same time on the subsequent day. This example, whilst trivial, illustrates in essence the philosophy of the approach.

The data for the west of the USA (Roehner 2007) exhibits a more regular behaviour than that for both London and Dublin. This, it is asserted, is what made the prediction slightly more effective in this case. The existence of cycles is not the key to the analysis. The moment at which a price peak occurs is irrelevant. What matters is the number of previous episodes and the fact that corresponding price peaks have the same *shape*.

Predictions were made towards the end of the up-going phase and focus on the price trajectory during the down-going phase. Once inflation adjusted prices have risen by a factor of two or three (as was the case in London), one can be sure that a speculative episode is underway. Taken alone this might be thought as insufficient to identify a peak, but from additional empirical observations the authors point to additional criteria that help identification of the tipping point:

- a sharp fall in the stock price of house construction companies

- a slow-down in the rate of price increases after a strong growth spurt

- an outright fall in house prices

However, the main point made by the authors is that speculative processes are so strong that all other variables (such as interest rates of levels of demand) become almost irrelevant. As a result, a model having no free parameters can be developed that enables the goal of predicting the price that occurs in the down phase. Note, however, that details of the shape of the price peaks may well be affected by other variables, including public policy. For example, the British government sought to prop up the market in 2005 and again in 2009–11, and the authors suggest this is responsible for the deviations from similarity observed in the latest speculative peak.

However, three successful predictions (UK, USA, and Ireland) suggest the presented approach may be on the right track, and in their latest publication, Richmond and Roehner (2012) look at current house prices in Beijing, Paris, and Seoul. The outcome, as China evolves through the present economic downturn, should be instructive and give more insight into the value of this empirical approach.

14

An introduction to physical economics

Can't Buy Me Love
  Lennon/McCartney

In Section 1.5 we commented briefly on the mechanical approach of the neoclassical theory of economics. For non-economists, such an approach is very difficult to understand. The idea that all future contracts and prices are somehow accounted for, or anticipated within present prices seems strange, if not fanciful. In Chapters 14 to 19 we shall present what we call *physical economics*, originally developed in various publications by Mimkes (2000, 2006a, 2006c, 2010a, 2010b, 2011, 2012) and Aruka and Mimkes (2011). The approach is based on two-dimensional calculus and a model first proposed by the French medical doctor François Quesnay (1694–1774). He may be thought of as one of the first economists who, inspired by the circulatory flow of human blood, considered a production process as a circuit.

14.1 Economic circuits

Quesnay analysed the simple case of a closed agricultural economy where each and every day farmers from households go to work into the fields. The farmers have capital, K, in the form of fields adjacent to the village. Each day, in return for their daily toil or work, W, they bring home consumer goods, G, from the fields. The process is captured in the closed *production circuit*, $\mathcal{P}$, shown in Figure 14.1.

The goods, G, are the reward for labour, W. This labour is directly related to the output of produce: a hard worker will pick many apples, other workers who may not work as hard pick fewer apples. Both work input and produce output are part of the same production circuit. In this picture, both work and the consumed produce may be measured in units of energy, but we will later choose to express these quantities in monetary units.

For survival and growth, the energy output of food from the fields must be larger than the energy input for work; the energy balance must be positive for each cycle. Only a positive net output leads to survival and growth; a continued negative net output would eventually lead to death. The length of the cycle depends on the system. People seek at least one meal each day. Farmers usually harvest in the fields once each year.

In these simple agricultural systems, trade took place by barter. Farmers bringing home potatoes from one field might exchange them with wheat, harvested by other

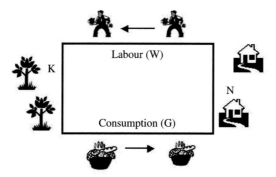

Figure 14.1 A model of a closed natural production circuit. Labourers from N households work in the fields (their capital, K). In return for their work, W, consumer goods, G, are brought back from the fields to the households.

farmers from another field. Today economic systems are more complex. Money, be it coins, paper notes, or in electronic form, is used by everyone as a store of value. This enables, for example, the baker to obtain apples from the grocer, when the grocer may not want bread from the baker.

The introduction of money creates a more complex economic circuit. In its most simplified form, the labour force from N households now works in industry, K. Moreover, unlike the fields of the farmer in the previous example, the workers do not generally own this industry. Nevertheless, goods, G, are created within the industry and acquired by the workers.

These goods, however, are no longer the direct reward for labour or work. Today the reward is money. The company pays wages, Y_H, for labour, W, and households pay costs, C_H, for goods and services. This generalization of the basic model of Quesnay dates back to the economist Irving Fisher (1867–1947) who recognized that the production cycle $\mathcal{P}$ of work and goods should be complemented by a second, *monetary circuit* $\mathcal{M}$, as illustrated in Figure 14.2. The monetary cycle is measured in a currency such as US dollars, \$, or euro, €, instead of energy. In Chapter 15 we shall formulate the equivalence of monetary and productive circuits in mathematical terms.

However, before we can do this, we need to develop some mathematical tools. Specifically we shall need to understand the basics of calculus in two dimensions.

14.2   Two-dimensional calculus for economic systems

Physics has long been a source of ideas for economics. Every investor would like to be able to predict the price of a stock in the same way as physicists predict the trajectory of a bullet fired from a gun. But while the flight of a bullet can be predicted *ex ante*, before it reaches its destination, one may not predict rates, prices, profits, or tax returns in this way, as we discussed in detail in the previous chapters. These quantities may only be stated *ex post*, after money has been earned. These non-predictable and predictable functions are called by economists *putty* and *clay* functions, respectively, terms introduced by Edmund Phelps (1963) to describe ideas by Leif Johansen (1959).

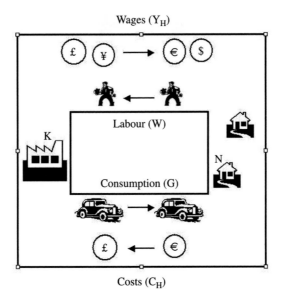

Figure 14.2 Diagram of an economic system that features both monetary and production circuits. Labour (W) is brought from (N) households to industry (K). In return, goods (G) are brought from industry to households. A second, monetary, or financial circuit ($\mathcal{M}$) quantifies the production $\mathcal{P}$ in monetary units. Industry pays wages (Y_H) for labour (W) and households pay costs (C_H) for goods G. K is the capital of the industry. We see in Chapter 15 that this picture leads to what may be regarded as an illustration of the fundamental law of economic theory ('Fisher's law').

Putty, an initially soft and deformable material, turns hard as it matures. In this vein, functions are called *putty functions*, if they are unpredictable; they are flexible *ex ante* and only fixed *ex post*. Company profits and income may only be *estimated* at the beginning of the financial year; only at the end of the year are profits or income fixed. Clay does not change its consistency over time (for constant water content). *Clay functions* are *predictable*; they are fixed both *ex ante* and *ex post*. The *production function*, which will be dealt with in Section 15.6, is one such example. Producers need to know before making an investment just how much capital and labour will be needed, and this should not change during the economic process.

In mathematics, clay and putty functions are termed *exact and inexact differentials* and play a key role in the framework of the calculus of two (or more) dimensions. In what follows we shall use these latter terms. The differentials were introduced into physics in order to develop thermodynamics in the late nineteenth century. Economists are well aware of these two types of functions, but seem not to have exploited them in order to develop a proper mathematical framework for economics. Maths books for economists include calculus in one dimension with exact differential forms only. However, as we have already intimated, many functions within economics are not exact in form and a two-dimensional calculus that also admits inexact forms is needed for a proper description. As we shall later see, this approach leads to a framework

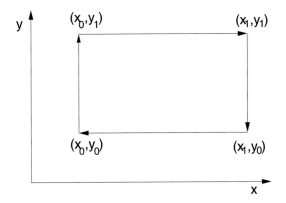

Figure 14.3 According to eqn (14.3), the closed Riemann integral of df is zero. The example shows the path $[x_0; y_0] \rightarrow [x_0; y_1] \rightarrow [x_1; y_1] \rightarrow [x_1; y_0] \rightarrow [x_0; y_0]$.

entirely consistent with the statistical framework developed earlier and applied to mathematical finance.

14.2.1   Exact differentials in two dimensions

An exact differential df of a function $f(x, y)$ has the form

$$df(x, y) = a(x, y)dx + b(x, y)dy = \frac{\partial f}{\partial x}dx + \frac{\partial f}{\partial y}dy, \tag{14.1}$$

i.e. the functions $a(x, y)$ and $b(x, y)$ fulfil the relation $\frac{\partial a}{\partial y} = \frac{\partial^2 f}{\partial y \partial x} = \frac{\partial^2 f}{\partial x \partial y} = \frac{\partial b}{\partial x}$ (see for example Flanders, 1990 and Cartan, 2006).

Integrals of exact differentials depend on the integration *limits*, but are independent on the *path* of integration. An integral around a closed loop in the $x - y$ plane is thus zero,

$$\oint df(x, y) = 0. \tag{14.2}$$

To see this, we divide the closed integral into two parts from $A = (x_0, y_0)$ to $B = (x_1, y_1)$ and back again, see Figure 14.3,

$$\oint df(x, y) = \int_A^B df + \int_B^A df = \int_A^B df - \int_A^B df = 0. \tag{14.3}$$

The integrals cancel, as they have the same limits but the opposite direction.

If the integrals of two exact differential forms are equivalent for any particular path, $\int_A^B df(x, y) = \int_A^B dg(x, y)$, then the functions $f(x, y)$ and $g(x, y)$ differ only by a constant of integration k, viz $f(x, y) = g(x, y) + k$.

Example 14.1 Consider the function $f(x, y) = x^4 y^7$. The mixed derivatives are identical, $f_{xy} = \frac{\partial^2 f}{\partial y \partial x} = 28x^3 y^6 = \frac{\partial^2 f}{\partial x \partial y} = f_{yx}$. Hence we obtain the exact differential $df(x, y) = 4x^3 y^7 dx + 7x^4 y^6 dy$.

The indefinite integral $f(x, y)$ of an exact differential form $df(x, y)$ may be found by computing the integral from (x_0, y_0) to (x, y) along the path $(x_0, y_0) \rightarrow (x, y_0) \rightarrow (x, y)$. Using $f(x, y) = x^4 y^7$ as above and setting $(x_0, y_0) = (0, 0)$, one obtains

$$\int_{x_0, y_0}^{x, y} df = \int_{x_0, y_0}^{x, y_0} df + \int_{x, y_0}^{x, y} df = \int_{x_0, y_0}^{x, y_0} 4x^3 y_0^7 dx + \int_{x, y_0}^{x, y} 7x^4 y^6 dy = 0 + \int_{x, y_0}^{x, y} 7x^4 y^6 dy = x^4 y^7.$$

The value of the integral is independent of the chosen path, e.g. taking (x_0, y_0) to (x, y) along the path $(x_0, y_0) \rightarrow (x_0, y) \rightarrow (x, y)$ results in

$$\int_{x_0, y_0}^{x, y} df = \int_{x_0, y_0}^{x_0, y} df + \int_{x_0, y}^{x, y} df = \int_{x_0, y_0}^{x_0, y} 7x_0^4 y^6 dy + \int_{x_0, y}^{x, y} 4x^3 y^7 dx = 0 + \int_{x_0, y}^{x, y} 4x^3 y^7 dx = x^4 y^7,$$

identical to the result obtained earlier.

Exercise 14.1 Select a path of your choice and show that the above integral remains unchanged.

14.2.2   Inexact differentials in two dimensions

For inexact differentials δg, with

$$\delta g(x, y) = a(x, y)dx + b(x, y)dy, \tag{14.4}$$

the partial derivatives $\frac{\partial a}{\partial y}$ and $\frac{\partial b}{\partial x}$ differ, i.e.

$$\frac{\partial a}{\partial y} \neq \frac{\partial b}{\partial x}. \tag{14.5}$$

The indefinite integral of $\delta g(x, y)$ may not be computed. The integrals depends both on the limits of integration A and B and on the path of integration,

$$g(x, y) = \int_{\text{path} A \rightarrow B} \delta g(x, y). \tag{14.6}$$

Furthermore, the closed integral of δg in the $x - y$ plane is not zero,

$$\oint \delta g(x, y) \neq 0. \tag{14.7}$$

The closed integral may be divided into two parts from $A = (x_0, y_0)$ to $B = (x_1, y_1)$ and back from B to A:

$$\oint \delta g(x,y) = \int_A^B \delta g + \int_B^A \delta g = \int_A^B \delta g - \int_A^B \delta g \neq 0. \tag{14.8}$$

The integrals along a closed loop in the $x - y$ plane will not cancel. They have the same limits in opposite directions, but the path of integration is different.

An inexact differential δg may be turned into an exact differential df by an integrating factor λ, e.g.

$$df(x,y) = (1/\lambda)\delta g. \tag{14.9}$$

In two dimensions this integrating factor λ always exists.

If the closed integrals of two inexact differentials (δg_1) and (δg_2) are equivalent, $\oint \delta g_1 = \oint \delta g_2$, then the difference between these two differentials is an exact differential, $\delta g_1 = \delta g_2 + df$, as $\oint df = 0$. The following examples illustrate the above discussion.

Example 14.2 A 'putty' inexact differential δg may be obtained by multiplying $df(x,y)$ of Example 14.1 by a factor x. In this case we have $\delta g(x,y) = xdf(x,y) = a(x,y)dx + b(x,y)dy = 4x^4y^7dx + 7x^5y^6dy$. We thus obtain $\partial a(x,y)/\partial y = 28x^4y^6$ and $\partial b(x,y)/\partial x = 35x^4y^6$, i.e. the mixed derivatives differ.

The inexact differential δg may be integrated from (x_0, y_0) to (x, y) along the path $(x_0, y_0) \rightarrow (x; y_0) \rightarrow (x; y)$ with $(x_0, y_0) = (0, 0)$, resulting in

$$\int_{x_0,y_0}^{x,y} \delta g = \int_{x_0,y_0}^{x,y_0} \delta g + \int_{x,y_0}^{x,y} \delta g = \int_{x_0,y_0}^{x,y_0} 4x^4y_0^7 dx + \int_{x,y_0}^{x,y} 7x^5y^6 dy = 0 + \int_{x,y_0}^{x,y} 7x^5y^6 dy = x^5y^7.$$

Let us now consider a different path with the same endpoints, e.g. $(x_0; y_0) \rightarrow (x_0; y) \rightarrow (x; y)$, which leads to

$$\int_{x_0,y_0}^{x,y} \delta g = \int_{x_0,y_0}^{x_0,y} \delta g + \int_{x_0,y}^{x,y} \delta g = \int_{x_0,y_0}^{x_0,y} 7x_0^5y^6 dy + \int_{x_0,y}^{x,y} 4x^4y^7 dx = 0 + \int_{x_0,y}^{x,y} 4x^4y^7 dx = \frac{4}{5}x^5y^7.$$

This example thus illustrates that the result of the integration of an inexact differential δg depends on the path of integration.

We can combine these two integration results to compute the integral along the closed path $(x_0, y_0) \rightarrow (x, y_0) \rightarrow (x, y) \rightarrow (x_0, y) \rightarrow (x_0, y_0)$ in the $x - y$ plane,

$$\oint \delta g(x,y) = x^5y^7 - \frac{4}{5}x^5y^7 \neq 0.$$

The integral along the closed path is non-zero for our inexact differential δg.

Example 14.3 The integrating factor. We had obtained δg by multiplying an exact differential $df(x,y)$ by a factor x. We may now integrate $df(x,y) = \delta g(x,y)/x$ to compute the indefinite integral $f(x,y)$. Accordingly, in this example $1/x$ is the integrating factor of the inexact differential δg.

Example 14.4 The closed 'Riemann' line integral. The closed line integral over an exact differential, which we may call 'Riemann' integral, is always zero (eqn (14.2)). This may be illustrated by:

    (a) a ring. After going around once, one is back to the beginning; nothing has changed.

    (b) a boat ride around Venice. After one circle the boat is back at the starting point, ready for another round. But it is also possible to take a boat across the island of Venice taking the 'Canal Grande', or several other canals.

    (c) a production cycle in which all harvested products are consumed without a surplus. The integral also resembles a monetary cycle without surplus, in which income equals costs.

Example 14.5 The closed 'Stokes' line integral. The closed line integral over an inexact differential, which we may call 'Stokes' integral, is always non-zero (eqn (14.7)). This may be illustrated by:

    (a) a spiral. After going around once, the spiral may look like in the beginning but we are not at the starting point, we have arrived at a higher level! Going in the opposite direction leads to lower levels of the spiral, but it is not possible to take short cuts.

    (b) the high and low pressure areas indicated on a weather map. In high pressure regions the winds run clockwise; in low pressure regions they run anti-clockwise. Winds can never take a short cut through the centre, the 'eye' of the vortex. At very low pressure, a tornado can lift heavy objects such as cars to a 'higher level'.

    (c) a production cycle with surplus. Work is the driving force in the centre of the production cycle. Profit is the driving force in the centre of the monetary cycle. After each cycle, the profit rises to a higher level. A deficit will change the direction and lead to lower levels of profit and eventually to debt and bankruptcy.

Example 14.6 Neoclassical model. Neoclassical economic theory interprets the monetary circuit as a *flow diagram*, where money flows from industry to households and there via household consumption eventually back to industry, as illustrated in Figure 14.4. This leads, however, to difficulties when trying to explain how industry can make money. In the example shown, it seems that industry receives back only €90 for every €100 invested so it must borrow another €10 from the bank at least just to break even.

This problem does not arise when interpreting production δP and money δM as *vortices*, resulting in non-zero values of integrals around closed circuits (see Example 14.5). The use of two-dimensional calculus is essential for this description.

With surplus, according to Example 14.5, the monetary circuit corresponds to a closed integral over an inexact differential (Stokes integral), resembling a spiral with many levels. There are no shortcuts possible to return (via banks) to the starting point. Another analogy is a high pressure area on the weather map. A high surplus

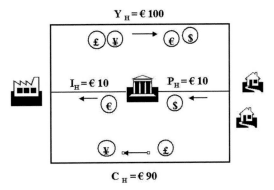

Figure 14.4 Illustration of the neoclassical interpretation of the monetary circuit as a *flow diagram*. Income Y_H flows from industry to households, consumption costs $-C_H$ flow from households back to industry. The surplus P_H flows to the bank and may be invested again by industry, I_H. In this picture, industry does not seem to be able to make money. This neoclassical model corresponds to the closed line integral over an exact differential ('Riemann' integral), which is only valid if no surplus is created, see Example 14.4.

leads to good prospects for the economic future; deficits correspond to low pressure areas, with bad expectations. A very large deficit in Greece at the time of writing leads to a heavy storm that will keep the other European countries busy for a long time. Additional deficit spending in Italy, Spain, and Portugal would lead to a 'bad weather' region in southern Europe and a good weather region in northern Europe.

15

Laws of physical economics

It was not by gold or by silver, but by labour, that all the wealth of the world was originally purchased; and its value, to those who possess it, and who want to exchange it for some new productions, is precisely equal to the quantity of labour which it can enable them to purchase or command.

Adam Smith, 'An Inquiry into the Nature and Causes of the Wealth of Nations'.

Having knowledge of two-dimensional calculus, as introduced in Section 14.2, we can return to the development of Fisher's approach to economic circuits (see Section 14.1). In particular we can now express the equivalence of the monetary ($\mathcal{M}$) and productive ($\mathcal{P}$) circuits in Figure 14.2 through the equality of two closed–line integrals over inexact differentials ('Stokes integrals'),

$$\oint \delta M = - \oint \delta P \neq 0, \tag{15.1}$$

and refer to this equation as 'Fisher's law'.

Both circuits are characterized by *inexact functions*, since neither the cash flow nor the production of a company can be predicted in advance. They may only be computed at the end of the financial period, be it after a week, a month, or a year.

The negative sign of the integral of production indicates that the production and monetary circuits flow in opposite directions. This equivalence, defined through the closed Stokes line integrals of eqn (15.1), forms the basis for all macroeconomic calculations that follow. Furthermore, since we are dealing with inexact differentials, the value of the closed integrals is generally non-zero.

15.1 The monetary circuit

We shall now examine the monetary circuit $\mathcal{M}$ via the two-dimensional calculus for inexact functions introduced in the previous chapter, see the illustration in Figure 15.1. The value of the closed integral $\oint \delta M$ is non-zero and equals the profit or surplus P_u arising from the production circuit. The suffix 'u' stands for a specific (inexact) production process,

$$\oint \delta M = P_u \tag{15.2}$$

(A negative value for P_u corresponds to a loss made.)

In the example of Figure 15.1, households gain $Y_H = €100$ during each cycle. If during the cycle, they also consume $C_H = €90$, then each household saves $P_H = €10$. As we have already remarked, the length of the cycle, which is determined by the

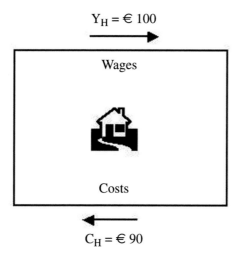

$$Y_H = € \; 100$$

Wages

Costs

$$C_H = € \; 90$$

Figure 15.1 Simple monetary circuit of modern households. Suppose each household earns $Y_H = €100$ in each cycle. If the households then spend $C_H = €90$, they save $P_H = Y_H - C_H = €10$ in each cycle.

economic partners (industry and households, managers, and unions) usually relates to a natural cycle (per hour, day, week, month, or year).

The closed Stokes line integral of eqn (15.2) is the basis of an economic action. Companies calculate their profit or surplus each year, for example the profit for 2012 may have been $P_{2012} = 10000$. This was as a result of specific production and sales strategies 'u'. We may say that monetary cycles δM are *driven* by profit or surplus P_u. Without profit there are no monetary circuits.

A positive surplus is needed for evolution and growth of economic, social, and biological systems. A deficit may be admitted, as long as reserves or loans are available. But deficits do lead to stress. Hunger makes people aggressive; declining stock markets make investors 'nervous'. At the time of writing we are seeing the stress within the Eurozone as a result of the lack of willingness to buy bonds of Eurozone countries that are not only in debt, but that are not generating growth. A continued deficit $P_u < 0$ eventually leads to bankruptcy in economic systems and death in biological systems.

Money is the outcome of the closed line integral of δM. Dimensionality has not yet been formally introduced to economics but now in this picture we may define 'money' as the common dimensionality of income, costs, surplus, profit, capital, production, etc. It may be measured in monetary units, i.e. currencies, such as US \$, European €, British £, Japanese yen.

The closed line integral of the inexact differential money δM (eqn (15.2)) may be split into two parts,

$$\oint \delta M = \int_A^B \delta M + \int_B^A \delta M = Y_u - C_u = P_u. \tag{15.3}$$

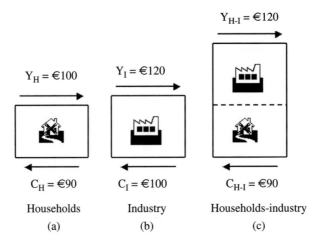

Figure 15.2 The line integral of monetary circuits is a balance that depends on the path of integration, on how money is earned and spent. This will be different for households (a) and industry (b). In the system households-industry (c) the balances of industry and households have been combined on top of each other. The total earning in the combined system (€30 in the shown example) is split into a fraction of one third for the households and two thirds for industry. We will return to this splitting of surplus in Section 18.2, when discussing economic growth.

Income, Y_u, and costs, C_u, are defined by open integrals over δM. The limits of integration extend from A to B, where the donor pays costs and from B to A, where workers receive income, i.e. $Y_u = \int_A^B \delta M$ and $-C_u = \int_B^A \delta M$. Income Y_u and costs C_u are path dependent integrals of δM and cannot be calculated before the money is paid. The surplus, P_u or profit, is measured as difference of income and costs per cycle and the dimension is also given in monetary units.

The monetary circuit is defined for specific systems. Figure 15.2 shows the monetary circuits for three systems: households, industries, and the combined system industry-households.

Households (Figure 15.2 (a)) may receive an income $Y_H = $ €100 from industry. Households will generally spend less than they earn in order to avoid debt. The costs may be $C_H = $ €90, which households spend on food and industrial goods. This way households may save $\Delta M_H = $ €10 in each cycle.

Industry (Figure 15.2 (b)) has wage costs, e.g. €100 per worker and cycle. In order to operate profitably, industry has to gain more than the wage costs, say €120 per worker and cycle. In this way industry gains $\Delta M_{In} = $ €20 per worker and cycle.

The income of industry producing a specific product cannot be paid by the workers alone, as they only earn €100 per cycle. The income will be paid by the retailer, who will again sell the product to customers at an even higher price. The customers must come from the wider community outside of the specific industry in order to be able to pay the high price of all the products. This may be done by selling products produced by countries where wage levels are relatively low, to consumers in countries where wage levels are higher, e.g. from Africa to the USA. In this way, labour may be exported and products imported, leading to a trade imbalance. Or one might trade the product for other products between companies or states. This keeps labour within the countries and leads to a more even trade balance.

In the *combined system* 'households-industry' (Figure 15.2(c)), the costs are $C_{In-H} =$ €90 for consumption of households, the income is $Y_{In-H} =$ €120. The income of households and the wage costs of industry cancel. The common profit is $\Delta M_{In-H} =$ €30 per worker and cycle.

The problem now arises as to how to split the surplus ΔM_{In-H} between industry and households. As surplus or profit are inexact or putty functions and cannot be calculated in advance, this split is usually negotiated later, between, for example, employers and employees. Game theory, as applied within the economics community, can offer some insight into such negotiations. We will return to the consequences of how the surplus is split when introducing a model for economic growth in Section 18.2 and also in Section 21.3, when discussing agent models of wealth exchange.

15.2   First law of physical economics

The equivalence of monetary (M) and productive (P) circuits defined by eqn (15.1) may be expressed in differential form, to result in what in the following we will call the *first law of economics*,

$$\delta M = dK - \delta P. \tag{15.4}$$

This equation is based on the laws of two-dimensional calculus for exact and inexact differential forms discussed in Section 14.2.2. If the closed integrals of two inexact forms δM and δP are equivalent, they may differ by an arbitrary exact differential form, dK. This exact differential dK must have the same dimension of money as income δM or profits and production δP. The negative sign of δP follows from the fact that labour has to be invested in order to make money. The exact differential form dK now refers to the *capital* of the production circuit. Equation (15.4) may be regarded as the financial balance of the economic system. It captures the following well-known facts.

- Profits or income δM depend on capital dK and labour δP.
- Income δM and production δP are inexact differentials and cannot be calculated 'ex ante', before the money has been earned.
- As the closed integral of dK is zero, eqn (15.4) leads to the conclusion that income δM is generated by labour δP, not by capital dK.

Adam Smith has articulated the third fact in the 'Wealth of Nations', cf. his quote at the beginning of this chapter.

So we see that two-dimensional calculus leads us to the important result that the equality $\oint \delta M = - \oint \delta P$ does *not* lead to the equivalence of δM and $-\delta P$, but leads to $\delta M = dK - \delta P$.

In eqn (15.4), labour or production, δP, refers to productivity, resulting from a labour force of many individuals. This number may, in times gone by, have been equivalent to the number of households. Today, both men and women are present in the (paid) workforce in almost equal numbers and the number of households is not always meaningful.

Table 15.1 Some important corresponding properties of economical and physical systems, together with the symbol, name, and unit in which the property is measured. In economics we also express energy in monetary units, as may be done by referring to the oil price.

Economics		unit	Physics		unit
M	money	currency	Q	heat	joules
K	capital	currency	E	energy	joules
P	production	currency	W	work	joules
λ	dither	currency	T	temperature	kelvin
S_e	economic entropy (production function)	*dimensionless*	S	physical entropy	joules/kelvin
π	economic pressure	currency	p	pressure	newton/m^2
A	freedom for action	*dimensionless*	V	volume	m^3

Equation (15.4) may be compared to the *first law of thermodynamics*, that relates heat Q, energy E, and work W via the relation

$$\delta Q = dE - \delta W. \tag{15.5}$$

We will in the following continue to develop a close correspondence between the functions/quantities of economics and the ones of thermodynamics, which is summarised in Table 15.1.

15.3   Second law of physical economics

Thermodynamic systems in equilibrium in the sense outlined in Section 1.4 may be characterized by a temperature, T. In mathematical terms, T is the integrating factor that relates our inexact differential form for the heat, δQ, to an exact differential form, dS. The new state function S is called *entropy* and the relation so formed is called the *second law of thermodynamics*,

$$\delta Q = TdS. \tag{15.6}$$

We now suppose by analogy that the inexact differential form of money δM may be related to an exact differential form dS_e via an integrating factor λ, (see Section 14.2.2), and write

$$\delta M = \lambda dS_e. \tag{15.7}$$

We call S_e *economic entropy* (Georgescu-Roegen, 1971) and will in the following refer to the above equation as the *second law of physical economics*. Unlike the physical entropy, S, which has dimensions of joules per kelvin, we introduce the economic entropy S_e as a dimensionless quantity. The integrating factor λ thus has the unit of a currency.[1]

In economics, S_e is usually called the *production function*. This function, we see from our definition, exists *ex ante* for the economic system and is a function of the way resources are allocated and structured within the economic system. We shall see later

---

[1]One could say that the introduction of the Boltzmann constant k_B was only necessary in physics because of the parallel existence of both concepts of temperature and energy before the link between heat and energy was properly understood.

that whilst the capital relates to the probability that any particular state of organization may occur, the production function/economic entropy captures the possibilities open to the system for internal organization of its resources. So, for example, in a manufacturing environment, S_e is determined by the number of workers and managers in the system, and the total number of ways they may organize themselves and the materials used in production. In trade, S_e is closely related to the variety and availability of products being offered. As we shall see, the state realized in practice is the most probable state and that for which the entropy is maximized. We shall return in Chapter 19 to a more detailed discussion of the entropy, which requires more concrete knowledge of the internal structure of specific systems.

15.4   Integrating factor or dither

The integrating factor λ has been termed the *dither* by E.T. Jaynes (1991). Whilst Keynes proposed the concept of animal spirit in an effort to account for erratic behaviour of individuals that led to economic booms, crashes, or recessions, the idea behind the dither is more general. It presumes to account for all the many independent decisions made by individuals involved in the economic process. In this sense it does not try, with Keynes, to separate so-called irrational behaviour from normal rational behaviour. Everything is supposed to be rational and, as we have already seen in Chapter 12, booms and crashes may arise as a result of rational behaviour.

For physical systems, we would at this point introduce the concept of *thermodynamic equilibrium*. Combining first and second laws of thermodynamics (eqns (15.5) and (15.6)), and eliminating the heat δQ, we obtain the physical equation

$$TdS = dE - \delta W. \tag{15.8}$$

It follows that the temperature is defined as $T = (\partial E/\partial S)_W$. The state of thermodynamic equilibrium is that state for which the temperature, T, is constant. Furthermore we can say that if a system A is in thermal equilibrium with system B and system B is in thermal equilibrium with system C, then system A is also in thermal equilibrium with system C. This is the so-called 'zeroth law of thermodynamics'. For a simple gaseous system, the temperature turns out to be proportional to the mean internal kinetic energy of the gas molecules. This is proportional to the mean square speed of the molecules as they move around the system and the function characterizing the distribution of molecular kinetic energies is the Maxwell-Boltzmann or Gaussian distribution (see the discussion in Section 5.2 and eqn (5.14)).

Turning to physical economics, we can combine the first and second law, eqns (15.4) and (15.7), to obtain

$$\lambda dS_e = dK - \delta P. \tag{15.9}$$

It thus follows that the dither λ is related to the entropy and capital in a similar manner, i.e.

$$\lambda = (\partial K/\partial S_e)_P. \tag{15.10}$$

We have no reason for believing the dither is constant across real economic systems. But we may infer from Figure 15.3 that periods do occur when only slow or quasi

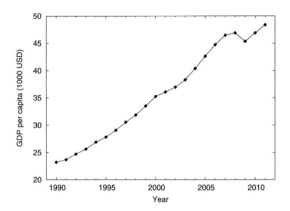

Figure 15.3 Evolution of the US gross domestic product (GDP) per capita between 1990 and 2011. The growth over the time period shown is approximately linear and of the order of $1000 per capita per annum. The USA as an economic system may be considered by analogy to a pot of water on a stove which is very slowly heating up. At every instant of time, t, the state of the system may be considered to be in equilibrium, corresponding to the instantaneous temperature, $T(t)$, or for economic systems, the GDP(t). (Data from 'List of countries by past and future GDP (nominal) per capita', International Monetary Fund World Economic Outlook (WEO) Database, April 2012 Edition.)

static changes take place, and at any moment of time we may assume it is constant. In Chapter 6 we discussed the approach of Bachelier to characterizing asset prices. He assumed that the volatility associated with the mean square fluctuations of (log-) price returns was constant and was thus led to a Gaussian distribution function for the (log-) price returns. We now know this assumption of constant volatility misses important features of the system behaviour, such as fat tail risk. The distribution of money across a society, whilst following a Boltzmann distribution for small values, also deviates from this Boltzmann distribution for large values of money, as will be discussed in Chapter 21. Furthermore, the total amount of money may change as a printing of money (so-called quantitative easing) by the authorities. Be this as it may, the theory of Bachelier has underpinned much of financial mathematics and economics. In the same spirit, we may expect that the introduction of Gaussian fluctuations via the dither will facilitate a theory of economics that is self-consistent and provides new insights beyond neoclassical economic theory.

The dither λ may be identified with the gross domestic product (GDP) per capita, which we may take as a proxy for the mean standard of living of a country, see Section 16.1. Empirically it is observed that whilst this may change over time, change usually occurs only slowly, as shown, for example, in Figure 15.3, for the evolution of US GDP over a time span of forty years.

15.5 Production laws

The first and second laws of economics, eqns (15.4) and (15.7), may now be combined into a single equation by eliminating money δM to obtain an equation for production output δP,

$$\delta P = dK - \lambda dS_e. \tag{15.11}$$

A production process involving the assembly of components, leading for example to a car, may be interpreted as follows. Production means arranging the disordered components of the car into the correct order; this is captured by a negative change in entropy, $-dS_e$. Entropy is a measure of disorder of a system, and $-dS_e$ stands for the reduction of disorder. The production process is identical to ordering. The factor λ, the standard of living in the producing country, determines the mean labour costs of ordering parts. The total labour costs are given by $-\lambda dS_e$; ordering a few parts with little change in entropy will cost little; assembling many parts of advanced products with a large reduction of entropy will cost much more. During the production process, an amount dK of capital may also be required ($dK < 0$), or may be produced ($dK > 0$). The outcome δP is measured in money and its value given in a specified currency.

Many governments today speak about becoming a knowledge economy. We may interpret this using the above ideas. We can say that if a system cannot only structure its labour but also gain an edge in the allocation of intelligence or brainpower, then this too must be captured within the entropic contribution to production. Brainpower or mental work effectively reorders ideas; teaching develops the minds of people young and old. Medical work reorders problems with the body. In the same vein, an edge in production may be gained through the use of advanced or new technology, arising from new ideas. The industrial revolution allowed the displacement of much labour with steam-powered machines, as machines follow the same (thermodynamic) laws as the human production laws, albeit displaying much more force. More recently we have seen computers displace labour and many machines developed in earlier times, due to their much better ordering capacity. In French the computer is called 'ordinateur', based on the word 'ordre': order.

The totality of production, however it is done, is captured by the inexact term δP, which may also be related to an exact differential form using another integrating factor, π. Thus we may write

$$\delta P = -\pi dA. \tag{15.12}$$

The equivalent thermodynamic relation concerns the work δW required to change the volume of gas by an amount dV against the pressure p. Thus

$$\delta W = -pdV. \tag{15.13}$$

Based on this analogy, we might say that the exact differential form dA is a measure of the *freedom for economic action* and call the function π, the *economic pressure*. In the following we shall treat A as dimensionless quantity; the economic pressure π then needs to have the unit of a currency for dimensional consistency of eqn (15.12).

Inserting eqn (15.12) into eqn (15.1) we obtain

$$\oint \delta M = -\oint \delta P = -\oint \pi dA = Y - C = \Delta M. \tag{15.14}$$

The economic pressure π of a household or an economy is closely related to the surplus or deficit ΔM at the end of the cycle, or at any other economic deadline! But it is not obvious how to compute or measure π, as A remains undefined. However, this quantity A has been introduced by the economist Robert Solow and used in his approach to neoclassical economic theory, where it is interpreted as *advancement of technology*. It is not our intention here to offer a discourse on neoclassical economic theory, but it is interesting to look briefly at the consequence of Fisher's ideas of economic circuits, as introduced in Section 14.1, when used within the neoclassical approach.

In Solow's neoclassical theory of economic growth, which, developed in 1956, remains the prevailing theory (Arnold, 2012), production output, Y, is determined by a *production function*, S_{nc}, that depends only on three quantities (see also Section 15.6). These are capital, K, the 'advancement of technology', A, and the size of the labour force, N (see for example D. Romer (1996) or Arnold (2012)),

$$Y = S_{nc}(K, A, N). \tag{15.15}$$

This is the neoclassical production law. A standard example of such a production function is the *Cobb-Douglas function* S_{CD}, given by

$$S_{CD} = AN^\alpha K^\beta, \tag{15.16}$$

where the exponents α and β are called *output elasticities* of labour and capital, respectively.

However, when viewed from the perspective of the theory we have developed so far, we can say that output Y is an inexact function. It is not known in advance. But the production function S_e is an exact function that depends only on the nature and structure of the system. So, without the concept of *dither* we have an inbuilt inconsistency. It is not possible to simply replace an exact function with an inexact function; one exception being the case when the integrating factor is constant. Only then does our theory reduce to a neoclassical economic theory without growth (cf. Section 18.1).

15.6 Production functions

The production function introduced above is a key element within economics. It captures crucial information about the production process. So, for example, the production function of a farm captures how much of each crop is planted. The production function of a company reveals quantities such as the number of people hired in different jobs and how many units of each product are produced.

Within the theory of physical economics developed here, we prefer to refer to the production function as economic entropy, S_e. It is obtained from combining the first and second law of economics (eqns (15.4) and (15.7)), respectively) and solving for dS_e,

$$dS_e = dK/\lambda - \delta P/\lambda = dK/\lambda + \pi dA/\lambda. \tag{15.17}$$

Here, we have used $\delta P = -\pi dA$ (eqn 15.12) to turn the inexact differential of production δP into the exact differential 'freedom for action' dA.

We see that the production function S_e of an economy depends on capital K, the dither λ, the equation of state or economic pressure π, and the freedom for economic action A. (For the correspondence with thermodynamic quantities see Table 15.1.)

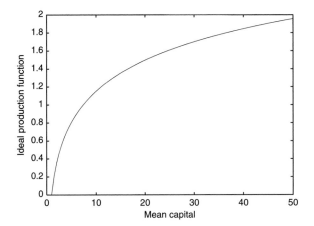

Figure 15.4 The ideal production function $\tilde{S}_e(\tilde{K}, A)$ of eqn (15.20), plotted as a function of mean capital $\tilde{K}$. We have set $A = 1$ and $\alpha = 1/2$ for illustrative purposes.

In order to progress, we require explicit expressions for K and π. We will show in Section 16.1 that for an 'ideal economic system', equivalent to the concept of an 'ideal gas' in physics, we obtain $K = \alpha\lambda N$ and $\pi = \lambda N/A$. Here N is the number of entities in the economic system, and α is a constant. Inserting these expressions into eqn (15.17) for dS_e we obtain

$$dS_e(K, A) = N\alpha dK/K + N dA/A. \tag{15.18}$$

If we introduce the production function *per capita*, $\overline{S}_e = S_e/N$ and the mean capital per capita $\overline{K} = K/N$, we may re-express eqn (15.18) as follows,

$$d\overline{S}_e(\tilde{K}, A) = \alpha d\overline{K}/\overline{K} + dA/A. \tag{15.19}$$

Hence by integration we obtain

$$\overline{S}_e(\overline{K}, A) = \ln(\overline{K}^{\alpha}A) + constant. \tag{15.20}$$

The *ideal production function* per capita, $\overline{S}_e(\overline{K}, A)$, is a logarithmic function, depending on the mean capital $\overline{K}$ and the 'freedom for economic action', A. This function is plotted as a function of $\overline{K}$ in Figure 15.4. Unlike in neoclassical theory, see eqn (15.16) and Figure 15.5, the ideal production function $\overline{S}_e(\overline{K}, A)$ is not zero for $\overline{K} \to 0$. According to the second law, $\delta Y = \lambda S_e$, income δY will go to zero, in the limit $\lambda S_e \to 0$.

Within neoclassical economics, A is considered to be a second and new production factor that takes account of 'advancement of technology'. Within the calculus theory developed here, this new production factor may be reinterpreted as a new economic variable, the 'freedom for economic action'.

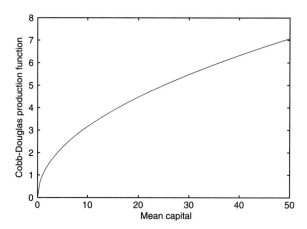

Figure 15.5 The Cobb-Douglas production function used in economics, see eqn (15.16), has a similar shape to the ideal production function of eqn (15.20), away from the origin. It is, however, *ad hoc* introduced, without a theoretical foundation. (Here we have set $\beta = 0.5$ and $A_t = 1$ and plot the Cobb-Douglas function per capita, with $\alpha = 1$.)

16
Markets

Was ist ein Einbruch in eine Bank gegen die Gründung einer Bank?
What is robbing a bank compared to founding a bank?
    Bertolt Brecht, 'Die Dreigroschenoper/The Threepenny Opera'.

In physics it has been useful to explore approximate models of reality. One such model is the *ideal gas*. In this model, N gas molecules move around freely within a volume V at pressure p and temperature T. The internal energy E is given by $E = n_f N k_B T$, where k_B is the Boltzmann constant and n_f is related to the number of degrees of freedom of the atomic elements making up the system ($n_f = 3/2$ for a monatomic gas). Volume and pressure are also related in a simple manner, viz the equation of state $pV = N k_B T$. At this point we notice that the temperature T is proportional to E/N or the mean kinetic energy per atom. In physics this corresponds to a special case of the equipartition theorem (see Section 5.2).

Based on the close relationship between physics and economics developed here, it is suggestive that we now consider an 'ideal' market or economic state.

16.1 Ideal markets

In analogy to the above equations for the ideal gas, and using the corresponding economical properties (see Table 15.1), the ideal economic state is described by the relations

$$K = \alpha \lambda N, \tag{16.1}$$

and

$$\pi = \lambda N / A. \tag{16.2}$$

where π and A are the economic pressure and freedom for action, respectively, as introduced in Section 15.5 and α is a dimensionless constant. The dither λ is now seen to be linearly proportional to K/N, the capital per capita. In economics the dither may be related to GDP per capita, the standard of living.

The first and second laws (eqns (15.4) and (15.7)) for income and profits, δM, in this ideal market are now

$$\delta M(\lambda, \pi) = dK(\lambda) - \delta P(\lambda, S_e) = \alpha N d\lambda - (\lambda N / \pi) d\pi, \tag{16.3}$$

and

$$\delta M(\lambda, \pi) = \lambda dS_e(\lambda, \pi). \tag{16.4}$$

Here we have used $\delta P = -\pi dA$ (eqn 15.12) and eqn (16.2) to turn the inexact differential of production δP into the exact differential 'economic pressure' $d\pi$.

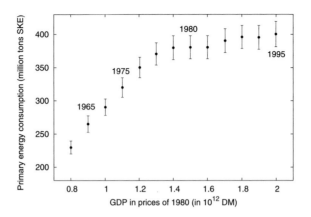

Figure 16.1 Economic development (GDP) and primary energy consumption in Germany from 1960 to 1995. Both may be taken as a proxy for the standard of living, and in the ideal economical system discussed here, also as proxy for the dither λ (see eqn (16.6)). (Data taken from Müller (2001). Primary energy in million tons of Steinkohleeinheiten (SKE) (1 kg SKE = 7.000 kcal = 29,3076 MJ = 8,141 kWh)).

At constant economic pressure, $d\pi = 0$, (e. g. without economic or social crisis, natural disaster, war, etc.) the second term on the RHS of eqn (16.3) is zero and income M depends only on the one variable λ. Hence

$$\delta M(\lambda) = \alpha N d\lambda. \tag{16.5}$$

As a result, the mean income (which may be measured as GDP per capita and be equated to the standard of living) is, for this ideal society, at constant pressure, also proportional to the dither λ,

$$m(\lambda) = M/N = const. + \alpha\lambda. \tag{16.6}$$

In modern societies, electric energy consumption is also an indicator of standard of living (Kümmel, 2011) and may thus serve as a proxy for the dither, λ. Figure 16.1 shows values for GDP and primary energy consumption of Germany, plotted as a function of time for the period from 1960 to 1995. At all times the GDP is proportional to energy consumption. A similar relation is displayed in Figure 16.2, which shows that there is a roughly linear relationship between GDP and energy consumption, based on data for 180 countries. As in our theory of physical economics (see Table 15.1), capital and energy may be seen as related quantities.

Income and profits at the ideal market depend on the two variables or production factors, dither λ and economic pressure π. According to eqns (16.3) and (16.4), the profit $\delta M(\lambda, S_e)$ may now be separated into two parts. The first contribution arises from a financial market where dS_e is zero. We shall consider such a market in Section 16.4. The second contribution arises from a productive market where λ is constant or varying only slowly, as discussed in the next section.

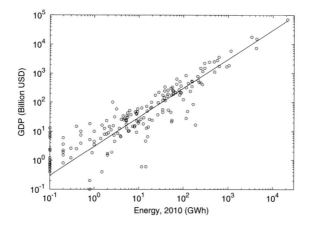

Figure 16.2 GDP for 180 countries in 2010 as a function of the total electricity production. The slope (dashed line) is equal to one, indicating a linear relationship between GDP and energy consumption. (Data: GDP: International Monetary Fund (2011). Electricity: BP workbook of historical data: gross electricity generation 1965–2010).

16.2 Slowly changing productive markets

Consider now using the theory above to determine how economic systems react to changes of the production factors λ and π. We first look at a production market for which λ is changing only slowly. Indeed, to a first approximation we suppose it to be constant in our theory, hence $d\lambda = 0$. The first and second laws now depend only on the one variable π.

According to eqn (16.4), profits are generated only through production, i.e. $dS \neq 0$. Suppose production takes place at $\lambda = \lambda_2$ whilst spending or consumption takes place at $\lambda = \lambda_1$. Income Y_u and costs or consumption C_u may be obtained by integration according to

$$Y_{\lambda_2} = \int \delta M_{\lambda_2} = \lambda_2 \int dS_e = \lambda_2 S_e + K_0, \tag{16.7}$$

and

$$C_{\lambda_1} = \int \delta M_{\lambda_1} = \lambda_1 \int dS_e = \lambda_1 S_e + K_0'. \tag{16.8}$$

where K_0 and K_0' are constants of integration, with the dimension of capital (money).

The profit $P_{u,\lambda}$ arises from the closed cycle of the two processes as

$$P_{u,\lambda} = \oint \delta M = Y_\lambda - C_\lambda = (\lambda_2 - \lambda_1)S_e. \tag{16.9}$$

Comparison with eqns (16.7) and (16.8) thus results in $K_0 = K_0'$.

We may eliminate S in eqns (16.7) and (16.8) to obtain the *consumption function* C_λ,

$$C_\lambda = Y_\lambda - S_\lambda = (1 - \lambda_1/\lambda_2)K_0 + (\lambda_1/\lambda_2)Y_\lambda. \tag{16.10}$$

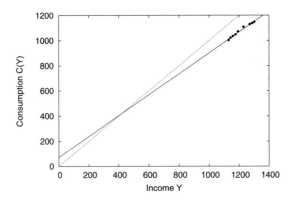

Figure 16.3 Consumption costs $C(Y)$ as a function of income Y for households at constant λ in Germany from 1993-2002 (expressed in prices of 1995). The function $C(Y)$ is well described by the linear relationship of eqn (16.10), resulting in the ratio $\lambda_1/\lambda_2 = 0.83$ and $C(Y = 0) = 71$, corresponding to $K_0 = 417$. The dotted line marks the 'no surplus' line, $C = Y$. (Data from 'Sachverständigenrat Wirtschaft', as reported on http://makroo.de/index.htm, see also Bofinger, 2003).

In neoclassical theory, C_λ corresponds to the so called *consumption function* $C(Y)$ which specifies how much consumers can spend as a function of the GDP (Arnold, 2012). It was developed by John Maynard Keynes in his book *The General Theory of Employment, Interest, and Money*, and is used to calculate the total amount of consumption in an economy.

From eqn (16.10) we see that C_λ emerges as a linear function of household income. In Figure 16.3 we plot empirical data for Germany for the consumption cost as a function of income Y and find that it is indeed well described a straight line. The constant slope, $\lambda_1/\lambda_2 = 0.83$, means that for every additional euro earned, 83 cents are spent. The intercept $C_{aut} = (1 - \lambda_1/\lambda_2)K_0$ is termed the *autonomous consumption function* in the economics literature and represents costs that need to be paid independent of the income.

Figure 16.4 shows the corresponding data for Canada over the period 1978 to 1998. The slope indicates a consumption rate of 67 cents per dollar; lower than that for Germany. This confirms again that a constant slope λ_1/λ_2 corresponds to a constant standard of living λ and a slowly changing (Canadian) market.

Countries or households with high income Y may have relatively high minimal costs C_{aut}. Countries or households with lower income will generally have lower minimal costs C_{aut}. In order to have a positive balance, where costs are lower than income, it is necessary that $C_{aut} < Y$. By reducing minimal costs, even countries with very low incomes can manage to retain a positive balance. This phenomenon may be observed in Figure 16.4. For years of relatively lower income (1978 to 1984), Canadians had lower minimal costs (cheaper rent for housing, cheaper cars for transportation etc.) than later years. But note these minimal costs C_{aut} vary only little compared to the change in income (Y).

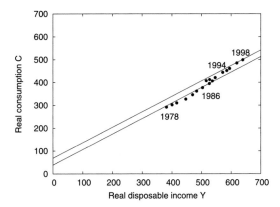

Figure 16.4 Canadian consumption function $C(Y)$ of households from 1978 to 1998, with data shown in billion of 1992 dollars. Both straight lines have slope 0.67; the upward shift in the data at the end of the 1980s corresponds to a change in the autonomous consumption function. (Empirical data taken from H.H. Chartrand's Compiler Press' Elemental Economics website, http://www.compilerpress.ca/ElementalEconomics/index.htm).

Quasi-static changes of the kind considered here, where λ is constant during economic change, correspond to isothermal changes in physical systems. The system is in thermal equilibrium and the equilibrium temperature is maintained by heat transfer with a larger environment. The same situation may be possible in economic systems. A slowly changing company structure may not necessarily change the dither too much. This will be more dependent on the demand for the products of a company. Lack of sales can quickly lead to lack of profit, depletion of company reserves, and possibly debt. This leads to nervousness and more rapid management activity.

Similarly, a slowly changing economy will also not change the dither of a country by much and empirically it can be observed that the foreign exchange rates can often be fairly stable over long periods of time. This was the position with many countries within the European Union during the latter part of the twentieth century and probably the reason politicians thought they were able to introduce a common currency. In effect they supposed fluctuations were Gaussian and did not exhibit power law tails. However, what is becoming clear at the time of writing is that the lack of control of spending within parts of this new common currency area during periods of low interest rates has given rise to unsustainable levels of debt within some of these countries. At the time of writing, EU politicians continue to seek financial bailouts to placate the creditors in the financial bond markets. However, this can only be a temporary fix for such high debts as seem to exist. The politicians argue that only with greater central control and coordination of individual member countries can the situation be stabilized. The theory proposed here would suggest that only by cutting expenditure and/or generating growth will an economic system avoid death.

Table 16.1 GDP, population and nominal GDP per capita for all countries in the Eurozone in 2011 (Data: Eurostat yearbook, 2011). Note that Ireland, which at the time of writing is in the midst of an economic crisis due to a bursting property bubble and a banking crisis, has a nominal GDP per capita slightly larger than that of Germany. For this reason, it is currently treated as 'special case' within Europe's bailout countries.

Member States	GDP 2011 (in billions of €)	Population (in millions)	nominal GDP per capita (in thousands of €)
Germany	2,571	81.4	31.4
France	1,997	65.3	30.6
Italy	1,580	60.3	26.0
Spain	1,073	46.0	23.3
Netherlands	602	16.6	36.1
Belgium	368	10.8	33.5
Austria	301	8.4	35.7
Greece	215	11.3	19.0
Finland	192	5.3	35.6
Portugal	171	10.6	16.0
Ireland	156	4.4	34.9
Slovakia	69	5.4	12.7
Luxembourg	42	0.5	82.7
Cyprus	18	0.8	20.6
Estonia	16	1.3	11.9
Malta	6	0.4	15.3
Eurozone	9,413	330.8	28.4
European Union	12,629	501.0	25.1

16.3   Common markets

A market across a group of countries may be discussed on the basis of ideal markets of countries at different standards of living. The mean standard of living is derived from the GDP per capita λ_i times the population N_i in each country:

$$\bar{\lambda} = \left(\sum \lambda_i N_i \right) / \left(\sum N_i \right). \tag{16.11}$$

In the equivalent physical picture, this may be compared to filling a bath tub with hot and cold water. If we fill it with ten litres of water at 60 °C and 15 litres of water at 20 °C, we obtain eventually 25 litres of water at 36 °C.

In a bath tub, hot and cold water mix by the free flow of heat. As a result, over time, the temperature of the total bath takes on the mean value. In the same way, one assumes that the flow of money between the group of countries, as a result of free trade, unimpeded by tariff barriers, will eventually lead to a common standard of living. Poor countries gain; rich countries lose.

Table 16.1 shows the GDP, the population, and the nominal GDP per capita for the European countries in the Eurozone in 2011. The standard of living in Greece or Portugal is much lower than the average in the Eurozone. Both countries gain in the Eurozone from money flows from Brussels for investments in infrastructure and from other support programs. Taxpayers in Germany and France lose some of their standard of living by paying more (to Brussels) than they receive.

But why should rich countries build a common market with poor countries, if they lose? The key point is that trade is the driving force here and as a result of trade, the parties involved increase their income from the profit of trade. As a result *both* parties usually become wealthier from the activity. Over time, then, there is a net growth in both economies engaged in trade. Providing there are no adverse shocks and this growth is a smooth process, the standard of living everywhere across the group of countries will steadily rise, albeit at different rates in the various countries to a new common value.

A union of countries always leads to a large economy playing a powerful political role. These large economies have developed in USA, China, and Europe and will continue to develop in India and South America. Single countries will have little chance to participate in their leadership.

A most important factor in pursuing an intact European Union is the stability of peace and freedom in this region. Before 1945, Europe was composed of a number of dictatorships: Hitler in Germany, Stalin in the Soviet Union, Mussolini in Italy, and Franco in Spain. This constellation lead to one of the most terrible wars in the world. After 1945, with the support of the Marshall plan of the USA, the European democracies started to prosper and by migration of workers turn more and more remaining dictatorships into democracies: Greece (1974), Spain (1976), Portugal (1977), the Eastern Block of Poland, East Germany, Czechoslovakia, Hungary, etc. (1989).

16.4 Rapidly changing financial markets

Rapid changes in economic systems, such as occur in financial markets and which are associated with rapid changes of the parameter λ, correspond to fast adiabatic changes in physical systems, where temperature may change, but heat cannot be exchanged, $(\delta Q = 0)$. Correspondingly for a stock market we have $\delta M = 0$. In this case, λ may be related to the level of price fluctuations which we have already seen can exhibit large clustered volatility where both the price and fluctuations of the price are rapidly changing.

During the complete cycle of an adiabatic process, the first and second laws, eqns (16.3) and (16.4), are one-dimensional and exact, and reduce to

$$\delta M(\lambda, \pi) = \alpha N d\lambda - (\lambda N/\pi)d\pi = 0, \tag{16.12}$$

and

$$\delta M(\lambda, \pi) = \lambda dS_e(\lambda, \pi) = 0. \tag{16.13}$$

The differential equation $d\pi/\pi = \alpha d\lambda/\lambda$ in our simple market now leads to a power law relationship between the economic pressure and the dither

$$\lambda/\lambda_0 = (\pi/\pi_0)^{1/\alpha}. \tag{16.14}$$

We may also calculate the 'freedom of action' A as a function of standard of living λ. With eqn (16.2) we obtain for eqn (16.3)

$$\delta M(\lambda, A) = \alpha N d\lambda + (\lambda N/A) dA = 0. \tag{16.15}$$

The differential equation $dA/A = -\alpha d\lambda/\lambda$ in our simple market leads again to a power law relationship between economic action A and the mean standard of living,

$$A(\lambda) = A_0(\lambda/\lambda_0)^{-\alpha}. \tag{16.16}$$

We have already seen powerlaw relationships in the distribution of log-price returns for financial markets. In physics, this process of alternating fast and slow motions is called a Carnot cycle, and is the basis of motors and generators. In economics, adiabatic changes are the basis for financial transactions which occur independently on a stock exchange, but also occur during production cycles. This Carnot process of economics systems will be discussed further in Chapter 18 and in Chapter 20 we shall show how generalized thermodynamics and statistical mechanics can be derived that allow for adiabatic fluctuations. For the moment we return to develop the consequences of our approach and move to examine trade.

17
A simple model of trade

Geld ist der Reliquie ähnlich. Man muss dran glauben sonst hat es keinen Wert.
Money resembles a relic. One has to believe in it, else it is worthless.

Christoph Süß, Morgen letzter Tag! Ich und Du und der Weltuntergang.

How can trade lead to prosperity? Two people meet in a market and trade commodities. Both become richer. How can this happen? Where does the wealth come from? Do traders rob each other? In this chapter we introduce a simple model of trade, together with a theory of supply and demand.

17.1 Trade in a two island world

Let us study the principle of trade by considering a simple two-island world. On one island, coconut island, suppose people harvest 100 coconuts per day. On another island, banana island, people collect 100 bananas each day. The people live off their crops. Imagine now one islander, a trader, who owns a boat.

The trader rows his boat between the two islands and makes his living exchanging bananas and coconuts. (Here we treat coconuts and bananas as having the same nourishment value.) The trader starts by borrowing one coconut, promising to repay it back later–with interest. With this coconut he rows to banana island where bananas are abundant and cheap, however, coconuts are rare and highly prized. So our trader finds he is able to obtain three bananas for one coconut. He now rows back to coconut island. Here coconuts are abundant and cheap but bananas are rare and expensive. He now trades his three bananas at a rate three coconuts for one banana. Now he owns nine coconuts.

By traveling between the two islands, the trader is buying at low prices and selling at high prices, thus making a profit on each trade. But with growing numbers of the other crop on each island, the exchange rate drops continuously until bananas and coconuts are nearly evenly distributed on both islands. At this point, no further profitable trading is possible. The trader returns home with his profit made before this equitable distribution is attained.

This in a nutshell is the solution to the old problem of trade: The people on the coconut and banana islands are not robbed. Nature is being 'robbed' as the environment provides nutrients and energy for the growing of the food and the traders and islanders share the spoils. This is same for all production processes: nature is the basis in farming, mining or any other economic activity. Humans, animals, plants live from exploiting natural resources.

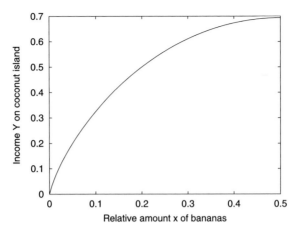

Figure 17.1 The trader's income on coconut island increases with the relative amount of bananas on the island according to eqn (17.2). There is no *profit* beyond equal number of bananas and coconuts, $x = 1/2$, since trade stops at this point. (We have set $\lambda N = 1$ for this example).

We can now calculate the trader's income on coconut island. The income, Y, results from selling bananas on coconut island. This may be calculated from the second law, eqn (15.7), as $\delta Y = \lambda dS_e$. The standard of living, λ, of the islands may be equated to the mean capital, which is measured by the mean number of bananas and coconuts per inhabitant. Since the total number N of coconuts and bananas on each island is nearly constant with time, the standard of living, λ, is constant, and the second law may be integrated to give $Y = \lambda S_e$.

In order to progress we need to compute S_e, based on the statistical interpretation of entropy, see Section (1.4). This results in the following expression for the *mixing entropy*,

$$S_e(x) = -N[x \ln(x) + (1 - x) \ln(1 - x)], \tag{17.1}$$

as will be derived in Section 19.2 (see also for example Mandl, 1998). Here $x = N_1/N$ is defined as the relative number of bananas, and thus $N_2/N = 1 - x =$ is the relative number of coconuts on coconut island. In microeconomics S_e corresponds to a *utility function*, see Section 19.2.

We can now readily compute the income, Y, of the trader on coconut island as

$$Y(x) = \lambda S_e(x) = -\lambda N[x \ln(x) + (1 - x) \ln(1 - x)]. \tag{17.2}$$

Leaving banana island, the income of the trader is measured in the number of bananas he brings to coconut island; leaving coconut island it is measured in coconuts. Figure 17.1 shows how this income (obtained from coconut island) varies as a function of the relative amount x of bananas.

At the beginning, at $x = 0$, the trader has no income. By rowing forth and back his income, the number of bananas he brings to coconut island, grows. At $x = 1/2$, i.e. when equal numbers of coconuts and bananas are reached, the income of the trader reaches a maximum. The price of coconuts and bananas is now equal and trade stops.

We may now calculate the *additional profit* $\Delta Y(x)$ that the trader makes for every banana that he brings to coconut island. Differentiation of eqn (17.2) with respect to x leads to,

$$\Delta Y = \frac{dY}{dx}\Delta x = -\lambda N[\ln x - \ln(1-x)]\Delta x = -\lambda N \ln(N_1/N_2)\Delta x. \qquad (17.3)$$

The additional profit ΔY per coconut (when $\Delta x = 1/N$) is then given by

$$\Delta Y = -\lambda \ln(N_1/N_2), \qquad (17.4)$$

i.e. it depends on the ratio of the numbers of coconuts (N_1) and bananas (N_2) on coconut island. The additional profit is high, if $N_1 \ll N_2$. The profit is zero if $N_1 = N_2$. We thus obtain

$$N_1/N_2 = \exp(-\Delta Y/\lambda), \qquad (17.5)$$

i.e. the ratio of coconuts and bananas (N_1/N_2) depends exponentially on the price difference, ΔY, relative to the standard of living, λ.

We conclude that:

- A price difference (ΔY) for two equivalent commodities leads to a difference in sales numbers (N_1/N_2).
- The production of two equivalent commodities in different numbers N_1, N_2 will lead to a price difference ΔY for these commodities.

17.2 Computation of functions for demand and supply

Demand and supply curves are used in economics to study what sets the price of traded quantities. We will, in the following, compute these functions based on the approach developed in the previous section.

Consider a simple market, at which only two different types of items are traded. The items, which in our model shall be of equivalent quality, are offered at prices y_1 and y_2 respectively, with $y_1 > y_2$. The majority (N_2) of buyers (demand) will choose the cheaper object (price y_2); only a minority (N_1) will choose the more expensive object at price y_1.

The resulting relations $\Delta y = y_1 - y_2 > 0$ and $N_1/N_2 < 1$ are consistent with eqn (17.5) which we may rewrite as

$$N_1 \exp(y_1/\lambda) = N_2 \exp(y_2/\lambda) = A. \qquad (17.6)$$

Since both y_1 and y_2 are independent variables, A must be a constant.

We thus obtain

$$N_D(y) = A \exp(-y/\lambda), \qquad (17.7)$$

which we may regard as probability based (stochastic) *demand function*. N_D is the number of buyers, or demand, at a price y and standard of living λ. A corresponds to the number of buyers when the price is zero and may be regarded as attractiveness of the product. Note that $N_D(y)$ is the Boltzmann distribution, that we already encountered several times (e.g. eqn (10.4)).

The number of sellers (supply) may be determined in a similar fashion by considering the price difference Δy in eqn (17.5) for producers/suppliers. A majority, N_1' will try to produce and sell the more expensive object (at price y_1), in order to earn more money. Only a minority, $N_2' < N_1'$ will try to produce and sell the cheaper object (at price y_2). We thus have the two relations, $\Delta y = y_1 - y_2 > 0$ and $N_1'/N_2' > 1$.

Consistency with eqn (17.5) requires $N_1'/N_2' = \exp[\Delta y/\lambda] = \exp[(y_1 - y_2)/\lambda]$. This can be rewritten as $N_1 \exp(-y_1/\lambda) = N_2 \exp(-y_2/\lambda) = const. = B$. The terms with index (1) are equal to the terms with index (2). As both indices are independent, B must be constant and may be regarded as equilibrium for supply. We thus arrive at

$$N_S^*(y) = B\exp(y/\lambda), \tag{17.8}$$

which may be regarded as stochastic supply function. N_S^* is the number of suppliers that will produce items at a price y and standard of living λ. B is the number of suppliers when the price is zero. Of course no company will usually produce commodities at zero price. Generally, there is a minimum price at which a commodity can be produced, or a minimum supply, N_0, at which profits can start. As a result, a more realistic supply function may take the form

$$N_S(y) = N_S^*(y) - N_0 = B\exp(y/\lambda) - N_0. \tag{17.9}$$

The *supply function* $N_S(y)$ has the property that the number of produced and supplied commodities $N_S(y)$ grows exponentially with profits (y), but profits will start only after a minimum number N_0 of commodities is sold.

At equilibrium, supply must equal demand, thus we obtain from eqns (17.7) and (17.9),

$$N_D(y) = A\exp(-y_e/\lambda) = B\exp(y_e/\lambda) - N_0 = N_S(y_e). \tag{17.10}$$

The solution for the equilibrium price y_e and the equilibrium number N_e of produced commodities may be obtained graphically or numerically. In Figure 17.2, supply and demand intersect at the equilibrium price y_e and equilibrium number N_e.

Note that eqn (17.10) results from a calculus-based economic theory and goes beyond standard economic theory, where generally only linear functions for supply and demand are discussed.

An interesting empirical measure of supply and demand may be found in data for the automobile market. Figure 17.3 shows the number of cars sold in different price categories. This corresponds to the demand function in Figure 17.2, which is met by the supply function at different equilibrium prices for different car types. The price of the very low cost new cars appears as an outlier, that does not follow the Boltzmann function. However, when adding in the sales of *second-hand* vehicles, the Boltzmann function applies to the entire price range. People on a low budget prefer to buy good used cars over new, but small cars, and this second-hand market completely dominates the low cost sales.

17.3   Inflation

In natural trade between banana and coconut islands, the crops are the capital stock of the islands. If the crop increases in the same way on both islands, the ratio of bananas

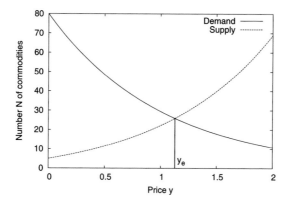

Figure 17.2 Equilibrium of demand (solid line) according to eqn (17.7) and supply (dashed line) in eqn (17.9) leads to an equilibrium price (y_e) and an equilibrium number (N_e) of sold commodities. (In the example shown we have set $A = 80$, $B = 10$, $N_0 = 5$, and $\lambda = 1$).

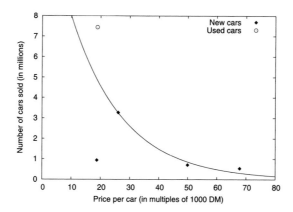

Figure 17.3 The automobile market in Germany in 1998 shows the number of sold new cars as a function of price. (Shown is the average price in four different tax categories that refer to different cubic capacities of the engine, less than 1400, 1400 to 2000, 2000 to 2400, and more than 2400 cc.) The lowest category is dominated by the used car market (square). The solid line is a fit to the demand function (Boltzmann distribution) of eqn (17.7). (Data: Statistik Kraftfahrzeug Bundesamt, Flensburg, 2000).

and coconuts will be constant and the prices in trade will stay unchanged. But both islands are now richer, they could feed more people.

If only the crop of bananas increased, banana island would be richer and could feed more people. But in trade, coconuts will become more expensive, as the price depends on the ratio of coconuts and bananas. This may be seen from Figure 17.2 as follows. As the supply curve moves upwards, the intersection point with the demand curve moves to the left, i.e. to lower prices. Thus bananas become cheaper, making the coconuts more expensive to purchase for the people on banana island.

There will be inflation in bananas. By this, the people on coconut island can buy more bananas, and even though their crop has not increased they can feed more people as well. By trade, local goods and also local wealth are spread to the trade partners.

The equivalence of productive and monetary cycles in modern markets (Chapter 15) requires a growth of money with growing production, if the prices are supposed to be stable. In economies, the growth of production in each cycle is based on the exploitation of resources. The corresponding growth of money requires the printing of more money by the central banks. In this way, the state participates at equal terms in economic growth.

The additional amount of money must correspond to the additional growth rate of production, which is, however, difficult to estimate in advance. Printing too much money will raise the price of goods and cause inflation. Printing not enough money will make money scarce and more expensive for investments. It is easier to adjust the production of money to the rate of inflation. Presently, central banks in Europe try to keep inflation down to about two percent.

But central banks are also urged by politics to offer cheap loans to investment banks in order to promote industrial production. This requires additional printing of money. This money will not affect prices as long as the money is repaid. But if a bank fails to repay—which is at present frequently the case—this will affect consumer prices. In this way, shoppers and taxpayers will have to pay for bank risks.

18

Production and economic growth

Der Physiker beobachtet Naturprozesse entweder dort, wo sie in der prägnantesten Form und von störenden Einflüssen mindest getrübt erscheinen, oder, wo möglich, macht er Experimente unter Bedingungen, welche den reinen Vorgang des Prozesses sichern. Was ich in diesem Werk zu erforschen habe, ist die kapitalistische Produktionsweise und die ihr entsprechenden Produktions- und Verkehrsverhältnisse.

The physicist either observes physical phenomena where they occur in their most typical form and most free from disturbing influence, or, wherever possible, he makes experiments under conditions that assure the occurrence of the phenomenon in its normality. In this work I have to examine the capitalist mode of production, and the conditions of production and exchange corresponding to that mode.

Karl Marx, Das Kapital. Kritik der politischen Ökonomie, Erster Band. Vorwort zur ersten Auflage, 1867.

In this chapter we shall adapt and exploit the concept of the Carnot cycle within our physical economics methodology to understand the mechanism of economic production and finance. Proposed by the French engineer Nicolas Carnot in 1824 and developed by Benoît Clapeyron, another French engineer, during the period 1830–50, the Carnot cycle is a theoretical concept that forms the basis for understanding the operation of engines and heat pumps or refrigerators. The cycle can be shown to be the most efficient route for the conversion of thermal energy into work (and so driving an engine) or creating and maintaining a temperature difference (which is the key to operating a heat pump or refrigerator).

18.1 Carnot cycle

Within our theoretical framework, every economic system exists in a particular state characterized by the capital K. As we remarked earlier, computing this function is not easy due to the difficulty in defining the economic pressure π for particular systems. However, we can make progress by considering an 'economic' Carnot cycle. If by some means we move our system through a series of different states and finally return it to its initial state, we may say an economic cycle has occurred.

Such a cycle may, by analogy with the ideas of Carnot, be called a *production cycle*. Within the economic framework it is clear that a complementary monetary cycle has also occurred and it follows from Fisher's law that this monetary cycle is in the opposite sense as the production cycle (recall that the integral of dK around a complete cycle is zero since dK is an exact differential, see also eqn (15.1)). Thus

$$\oint \delta M = - \oint \delta P \neq 0. \tag{18.1}$$

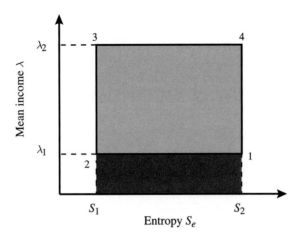

Figure 18.1 A generalized balance or monetary cycle. The lower dark shaded area is the amount of money taken out by the 'poor' reservoir (costs $C = \lambda_1 \Delta S_e$). The upper lighter shaded area is the amount of money (surplus or profit) generated by the system during a complete cycle of production. The amount of money injected from the 'rich' reservoir is the sum of the two (income $Y = \lambda_2 \Delta S_e$). The production process happens as the cycle moves clockwise around the loop, and the monetary cycle is computed from a counter-clockwise calculation.

According to the second law, such a cycle may be performed within a region of (λ, S_e) space, where λ is the dither (interpreted here as before as 'standard of living') and S_e the entropy or production function of the economic system,

$$\oint \delta M = \oint (\lambda dS_e) = \Delta \lambda \cdot \Delta S = Y - C. \tag{18.2}$$

The Carnot cycle is a closed line integral of the second law and is equivalent to the balance of income Y and costs C. In the (λ, S_e) plane it results in movement around a *rectangle*, as shown in Figure 18.1. The integral is easy to calculate. Along the horizontal elements λ is constant; along the vertical elements S_e is constant. This area within the rectangle is the maximum possible outcome between the values λ_1 and λ_2.

Entropy may be regarded as a measure of disorder of a system (of goods and products, money or other system elements). A negative entropy change, $dS_e < 0$ means reduction of entropy, for example by collecting system elements. A positive change, $dS_e > 0$ means production or distribution of elements.

The production process over a cycle always leads to an overall increase in entropy since the area under the curve corresponds to products being sold to customers with $\lambda = \lambda_2$ as a result of products being bought from poorer workers with $\lambda = \lambda_1 < \lambda_2$. For a counter-clockwise monetary cycle, the area under the upper portion is the income, Y, collected during the cycle from customers while the area under the lower portion yields the costs, C, paid to the workers. The area inside the closed circuit, which is the difference, represents the surplus (Δ M = Y – C) or profit of the production process.

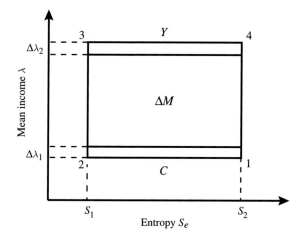

Figure 18.2 The mechanism of economic growth. The surplus $\Delta M = \Delta M_1 + \Delta M_2$ with $\Delta M_1 = \Delta \lambda_1 \Delta S_e$ and $\Delta M_2 = (\Delta \lambda_2 \Delta S_e$ may be distributed to both levels, leading to new levels $(\lambda_1 + \Delta \lambda_1)$ and $(\lambda_2 + \Delta \lambda_2)$, at which the next Carnot cycle will continue.

After each cycle the surplus ΔM in Figure 18.1 may be split into two parts,

$$\Delta M = \Delta M_1 + \Delta M_2, \tag{18.3}$$

and then distributed to the two parties, for example exporters-importers or workers-customers at (λ_1) and (λ_2), as shown in Figure 18.2. The profits ΔM_1 and ΔM_2 lead to new levels $(\lambda_1 + \Delta \lambda_1)$ and $(\lambda_2 + \Delta \lambda_2)$ in the next cycle. In this way, economic *growth* is obtained in each cycle, as will be discussed in detail in Section 18.2.

Let us now look at the three examples, farmers market, international trade, and banking to illustrate the Carnot process. The numbers in these examples refer to the numbering system in Figure 18.1.

Example 18.1 Farmers market
Apples are grown on a farm at λ_1 and sold in a market at λ_2.

Production cycle (δP) (clockwise direction)
$1 \rightarrow 2$: Apples are collected $(\Delta S_e < 0)$ in a plantation at low wage level (λ_1).
$2 \rightarrow 3$: Apples are brought $(\Delta S_e = 0)$ from the plantation (λ_1) to the market (λ_2).
$3 \rightarrow 4$: Apples are distributed $(\Delta S_e > 0)$to customers at high price level (λ_2).
$4 \rightarrow 1$: Fertilizers from waste are brought $(\Delta S_e = 0)$ back to the fields $(\lambda_2 \rightarrow \lambda_1)$.

Monetary cycle (δM) (anti-clockwise direction)
$4 \rightarrow 3$: Farmer collects $(\Delta S_e < 0)$ money from customers at high price level (λ_2).
$3 \rightarrow 2$: Money transfers $(\Delta S_e = 0)$ from the market (λ_2) to the plantation (λ_1).
$2 \rightarrow 1$: The farmer pays $(\Delta S_e > 0)$ low wages to plantation workers (λ_1).
$1 \rightarrow 4$: Workers transfer $(\Delta S_e = 0)$ money to the market by shopping $(\lambda_1 \rightarrow \lambda_2)$.

Example 18.2 International production and trade

Coal mined in South Africa (SA) at low price level (λ_1) is sold to European companies (EU) at high price level (λ_2) and traded for European goods.

Production cycle (δP) (clockwise direction)

$1 \rightarrow 2$: Coal collected ($\Delta S_e < 0$) from mines in SA at low wage level (λ_1).

$2 \rightarrow 3$: Coal shipped ($\Delta S_e = 0$) from SA (λ_1) to the EU (λ_2).

$3 \rightarrow 4$: Coal distributed ($\Delta S_e > 0$) to EU customers at higher price level (λ_2).

$4 \rightarrow 1$: EU products are brought ($\Delta S_e = 0$) back to SA ($\lambda_2 \rightarrow \lambda_1$).

Monetary cycle(δM)(anti-clockwise direction)

$4 \rightarrow 3$: EU importers collect ($\Delta S_e < 0$) money from EU customers at high price level (λ_2).

$3 \rightarrow 2$: EU importers transfer ($\Delta S_e = 0$) money from EU (λ_2) to SA (λ_1) exporters.

$2 \rightarrow 1$: SA exporters pay ($\Delta S_e > 0$) low wages (λ_1) to SA workers.

$1 \rightarrow 4$: SA workers transfer ($\Delta S_e = 0$) money back ($\lambda_1 \rightarrow \lambda_2$) for EU products.

Example 18.3 Banking

Savings and investments require different rates of interest.

Production cycle of capital (δP) (clockwise direction)

$1 \rightarrow 2$: Capital is collected ($\Delta S_e < 0$) from savers at low interest level (λ_1).

$2 \rightarrow 3$: Capital is sent within bank ($\Delta S_e = 0$) from savings accounts (λ_1) to investment managers (λ_2).

$3 \rightarrow 4$: Capital is loaned out ($\Delta S_e > 0$) at high interest level (λ_2).

$4 \rightarrow 1$: Capital is sent ($\Delta S_e = 0$) from the investment (λ_2) to the savings desk (λ_1).

Monetary cycle of interest(δM)(anti-clockwise direction)

$4 \rightarrow 3$: Banks collect ($\Delta S_e < 0$) interest from investors at high interest level (λ_2).

$3 \rightarrow 2$: Banks send ($\Delta S_e = 0$) interest from investment managers to savings accounts (λ_1).

$2 \rightarrow 1$: Banks pay ($\Delta S_e > 0$) low interest rates (λ_1) to savers.

$1 \rightarrow 4$: Banks return ($\Delta S_e = 0$) from the savings (λ_1) to the investment desk (λ_2).

The surplus arising from a Carnot cycle in banking may be reinterpreted in terms of risk. Suppose we require a fixed profit or surplus. Since $\Delta S_e \Delta \lambda = \Delta M$, this may be obtained by having a low interest differential and high entropy or a high interest differential and low entropy. Since we know from Chapter 11.2 that returns are proportional to risk, then the risk may be equated to the inverse entropy, $1/\Delta S$. Alternatively, we may consider ΔS as a safety index. According to the definition of entropy (see Chapters 1.4 and 19.1), we may define the safety of a portfolio, divided into n assets with weightings p_k, as

$$\Delta S_e(n) = -\sum_{k=1}^{n} p_k \ln p_k. \tag{18.4}$$

For only one asset and thus $p_1 = 1$, the safety index is $\Delta S_e(1) = 0$ and the risk is infinite. Hence the proverb 'do not carry all eggs in one basket'. Security will grow with the number of different assets, but returns will diminish.

The mechanism of economic production and finance in Figure 18.1 applies to all systems, to people at work, to small business or big companies, to single persons or countries and economies: *Production and selling of goods always requires two different price levels, λ_1 and λ_2!*

In economies we have different standards of living. Production demands that workers receive a low wage (λ_1), whereas consumers pay out at a higher level (λ_2). Within companies, we have low income labour (λ_1) and higher paid capitalists (λ_2). In international production we often find two standards of living in the so-called 'third' and 'first' worlds. In banking we have high and low interest levels, λ_1 and λ_2. In societies we always find both rich and poor (see also Chapter 21). Similarly, for a thermodynamic Carnot engine, two different temperature levels are required. For an engine, fuel oil is usually needed, which ultimately derives from solar activity. In the same way, production processes ultimately utilize energy from the sun.

18.2 Economic growth

Let us now see how the above described Carnot process may be used to describe how economies may evolve in time. The key point to bear in mind is that the level of λ for the two groups interacting in the process differs by a small amount $\Delta\lambda$, so that over the cycle, a surplus or profit can be generated. This surplus is then split between the two groups, according to some agreed formula. (An example of this is illustrated in Figure 15.2). This is usually the result of a negotiation that may be reviewed from time-to-time. Within a company, the workers might join a union that may negotiate with the owners. Trade agreements between countries are negotiated within meetings of world trade conferences, such as GATT (general agreements on tariffs and trade).

Suppose then Group 1 takes a fraction p of the surplus and thus Group 2 a fraction $1 - p$. Over a short time, dt, it follows that the income changes dY_1 and dY_2 of the two groups are given by

$$dY_1(t) = p(Y_2 - Y_1)dt,$$
$$dY_2(t) = (1 - p)(Y_2 - Y_1)dt. \tag{18.5}$$

Exercise 18.1 Show that the solutions of these equations are given by

$$Y_1(t) = Y_{10} + p(Y_{20} - Y_{10})(\exp[(1 - 2p)t] - 1)/(1 - 2p),$$
$$Y_2(t) = Y_{20} + (1 - p)(Y_{20} - Y_{10})(\exp[(1 - 2p)t] - 1)/(1 - 2p), \tag{18.6}$$

where $Y_{10} = Y_1(t = 0)$ and $Y_{20} = Y_2(t = 0)$ are the initial values.

Let us now show how Y_1 and Y_2 vary with time, depending on how the profit is split between the two groups, i.e. on the value of p.

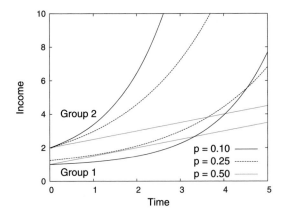

Figure 18.3 The development of income $Y_1(t)$ and $Y_2(t)$ of two interdependent economic systems starting at $Y_{10} = 1$ and $Y_{20} = 2$, as obtained from eqn (18.6) for the parameter $p = 0.1, 0.25$ and 0.5. In the long run, the income $Y_1(t)$ of the initially poorer group will grow less quickly than that of the initially richer group, $Y_2(t)$, to whom it is coupled.

Exponential growth for $0 < p < 0.5$

If $p = 0$ then all the profit goes to the richer party. The income of Group 2 will grow exponentially, while the income of Group 1 stays constant.

However, if p is greater than zero, but still less than 0.5, say $p = 0.10$, both parties grow exponentially, as shown in Figure 18.3. This might describe the case for Japan and Germany after World War II, when both economies were strongly dependent on the US and were growing exponentially. Another more recent example may be found in the trade relation between US and China, see Figure 18.4.

For the case of $p = 0.25$, also shown in Figure 18.3, where the poorer party Y_1 receives 25% of the profit, the income of the richer Group, 2, still grows exponentially. But as the poorer side is linked to a less rapidly growing rich side, it actually grows less over the long-term, than for the situation where it receives 10% of the profit ($p = 0.1$).

A rapid rise in wages may thus weaken the economy and lead to lower wages in the long run. A lower increase in wages will lead to a greater return for both industry and workers over the long-term. This suggests that workers, as well as their managers, should be patient with pay rises.

The fair deal and linear growth for $p = 0.5$

If $p = 0.5$, and the profit is split evenly between the two parties, a surprising result is observed. An even split between the two parties seems to be a fair deal, but this 'fair' deal leads only to linear growth for both parties. Linear growth is usually of no good to anyone engaged in wage negotiations. However, for a short time, such linear growth is actually more profitable for the poorer party, as indicated in Figure 18.3. And as wage negotiations are only valid for short times, linear growth could in fact be an attractive option for the poorer side. The same is true for managers with short time contracts. A manager who may be keen to move on will generally take the option

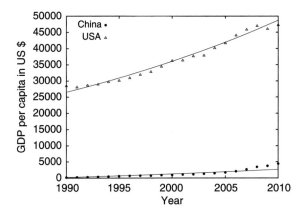

Figure 18.4 Economic growth of the two party system China and USA. The data corresponds to economic growth in Figure 18.3 at $p < 0.5$, where the growth rates are still super-linear. The large difference in GDP per capita is the basis of the strong economic motor China–USA. The solid lines correspond to calculations according to eqn (18.6) for $p = 0.1$ and $t = 0.035n$, where n refers to the year. (Data: International Monetary Fund (IMF), 2008).

of high, short-time growth with high personal benefits, rather than invest in future exponential growth.

Convergence for $0.5 < p \leq 1$

If $p > 0.5$, eqn (18.6) leads to decreasing efficiency of the system, as measured by the efficiency

$$\eta(t) = \frac{Y_2(t) - Y_1(t)}{Y_1(t)}, \tag{18.7}$$

which declines with time. Now Y_1 trails behind an ever slower increasing Y_2 and both Y_1 and Y_2 finally reach a common plateau (see Figure 18.5). After World War II, Japan built many production plants, allowing the country to share in the economic growth, previously the province of the USA. As a result, the rate of growth in both countries began to slow down until both were converging to roughly the same level (see Figure 18.6).

A similar development was expected in the relationship between West and East Germany after reunion in 1990. Figure 18.7 shows the GDP per capita for the eastern and the western part of Germany after reunion, over a period of time. The calculated lines show the expected GDP per capita to converge after ten years. In 1998, East Germany reached about 80% of the mean income in West Germany (Fründ, 2002). The factor p may be estimated from the compensation payments of states in the former West Germany to those in the former East. More recent data, however, indicate that not even by 2012 will the two parts of the country have converged in GDP per capita. The reason is the continuing difference in productivity between the two parts of the country, and the insufficient and decreasing support of the western states for the eastern states of Germany.

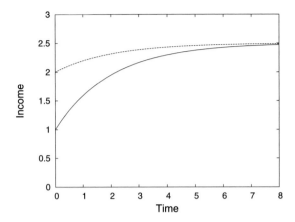

Figure 18.5 The development of income of two interdependent economic systems starting at $Y_{10} = 1$ and $Y_{20} = 2$. At high values of profit for the poor side, $p = 0.75$, economic growth asymptotes to $Y_1(t) = Y_2(t) \to 2.5$ for $t \to \infty$ (see eqn (18.6)).

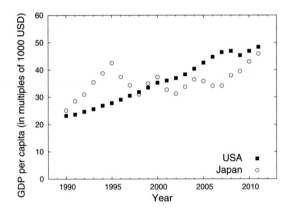

Figure 18.6 Development of mean income (GDP per capita in multiples of 1000 US dollar) of USA and Japan between 1990 and 2011. The variations of the Japan GDP per capita reflects the variations of the exchange rates of US dollar and Japanese yen between 1990 (1USD=140Y) and 2010 (1USD=100Y). According to central banks official exchange rates, the yen was valued too low between 1990 and 2000, and too high between 2000 and 2010. (Data from 'List of countries by past and future GDP (nominal) per capita', International Monetary Fund World Economic Outlook (WEO) Database, April 2012 Edition.)

Convergence for $p > 1$

If $p = 1$ and all the surplus goes to the poor side, the income of the poor party soon reaches the constant income of the rich party, and the efficiency of the system, as defined by eqn (18.7), is zero. If $p = 1.2$, we have the situation as illustrated in Figure 18.8. If more than 100% of the profit goes to the poor party, the growth of Y_2 is negative and ultimately both parties attain the same level.

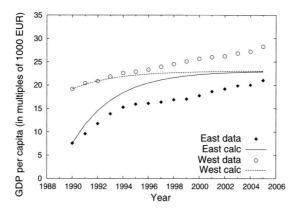

Figure 18.7 Development of mean income in former West and East Germany between 1990 and 2005. The solid line corresponds to $p = 0.8$ in our growth model, as estimated from compensation payments. It shows a convergence which is not in line with the data. (Data: Statistisches Jahrbuch der Bundesrepublik Deutschland 2012).

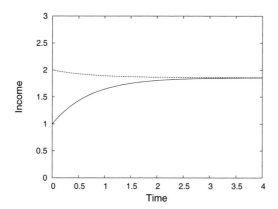

Figure 18.8 The development of the income of two interdependent economic systems, starting at $Y_{10} = 1$ and $Y_{20} = 2$ for $p = 1.2$. Both economies will eventually converge to an income below the starting value of the rich system.

Exploitation: $p < 0$

The case $p = -0.25$ is illustrated in Figure 18.9 and might best be described as exploitation. Now the poor side Y_1 suffers only losses and will eventually go bankrupt. The richer party, Y_2, however, will grow exponentially.

A similar situation was observed in Argentina in 2000 when many rich people transferred assets to banks in the USA (see data in Figure 18.10). Due to the large amounts of capital transferred to the USA, the mean income in Argentina was much reduced. The small ongoing exponential increase in the USA GDP as compared with the strong increase in Figure 18.9 may be considered to be due to trade deficits with other important trading partners, such as Japan. In Figure 18.10 the exponential

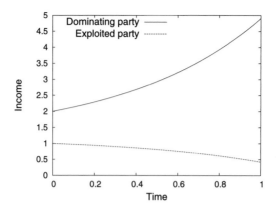

Figure 18.9 For $p < 0$, the income of the dominating party grows exponentially while the exploited party falters. The calculation shown is for $p = -1/4$.

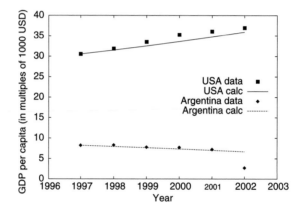

Figure 18.10 The development of mean income of USA and Argentina between 1997 and 2001, which grows by about 10% in the USA and drops by 10% in Argentina. (Data: World Factbook, USA, 2004: Ministerio de Economa, Argentina, 2004). The solid lines correspond to a calculation with $p = -0.4$.

trends for Y_1 and Y_2 are not very pronounced, but do suggest that negative values of p are possible.

Another example could be the current situation in Europe. The economies of both Greece and Spain are declining as a result of severe government imposed austerity, and capital flight, as rich people move their money out of the country. The economy of trading partner Germany is, however, continuing to grow with a positive trade surplus over trading partners Greece and Spain.

The above examples illustrate that our simple Carnot process, applied to the surplus distribution for two interacting groups, is able to illustrate various growth scenarios as found in economic data. In the real world, the details of this distribution results from lengthy discussions and may change over time. However, our examples illustrate the key principle of profit generation and economic growth of businesses and economies.

In Chapter 21, which deals with the distribution of wealth in society, we will return to eqn (18.5) as an exchange rule for a model of interacting agents (Section 21.3). Our numerical simulations show that using our 'Carnot rule', we are able to obtain power-law like wealth distributions, similar to those as seen in empirical data. For the remainder of this chapter, however, we shall continue with our discussion of the production process, as arising from the Carnot cycle.

18.3   Stable economies

When an automobile is started up and accelerated, the engine warms up and a temperature gradient builds up between the hot inside and the cool outside. This phase of acceleration corresponds to the phase of economic growth of an economy. After some time, the motor reaches a steady speed and the Carnot motor is operating with a constant temperature gradient between inside and outside. At this point, the rate of energy generated by the motor is constant and used only to overcome the frictional forces of air on the automobile body and the road on the tyres.

So-called emerging economies (such as at the moment China and India) enjoy high growth rates; they are still in the process of acceleration. Mature economies such as the USA, Japan, and Germany have only modest growth rates; they are almost stable. In these countries the income Y or GDP is used mainly to maintain the existing standard of living, $Y - C \simeq 0$. The income generated within the economy is consumed essentially by wages, taxes, interest on investment loans, and maintenance of infrastructure which may include the military, roads, hospitals, and schools, etc.

A modern automobile essentially runs by itself and needs only a few periodic check-ups. But after long times of operation it may get too hot, the cooling system may fail or some other technical problem may occur. Then an expert mechanic is required.

In the same way, a well structured economy may run by itself, needing little control by government. But over the long-term problems arise. The balance of supply and demand may get out of kilter, companies may have grown and, via take-overs, limited competition, leading to high prices. Governments themselves may have formed monopolies, see for example the health-care system in the UK, again leading to monopoly pricing and taxation policies that reduce the opportunities for growth of the private sector which comprises the economic Carnot engine. It is therefore important that government has access to experts who can advise properly on the steering of state expenditure and the fair regulation of the private sector.

18.4   GDP, energy, pollution

We have seen that the Carnot process leads us to an understanding of the mechanism of production and the need for energy. The relationship between economic processes and energy consumption is illustrated by the data shown in Figure 18.11 for 126 countries. GDP per capita and energy consumption are strictly proportional! The higher the performance of a motor, the more fuel is required, the higher the standard of living, the more energy seems to be consumed. Of course, some motors run more efficiently

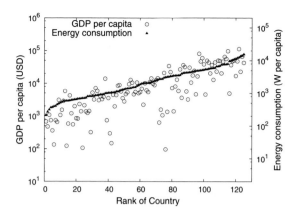

Figure 18.11 GDP per capital (2010) for 126 countries, data from International Monetary Fund (2010–11) and total energy consumption per capita, ranked by energy consumption per capita, as published by the World Resources Institute for the year 2003. (Data: US Energy Information Administration: Governmental Consumption Reports 2010).

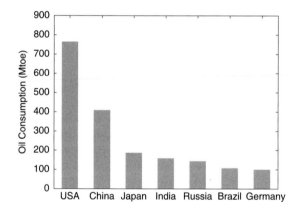

Figure 18.12 Ranking of the seven biggest oil consuming countries in 2011. The data is given in Mtoe, with one tonne of oil equivalent, toe, given by 1 toe = 41.87 gigajoules. (Data from Global Energy Statistical Yearbook 2012).

than others, and some countries are more efficient in production than others. This leads to the variations of the data for production shown in the figure.

The close relationship of GDP and energy consumption also points to the problems of pollution and arguably climate change. Industrial production, we have seen, corresponds to a reduction of entropy in the production system (company). However, by using energy for production and finally when the product is discarded, we have entropy production outside of the production system (company), which manifests itself as pollution. The richest nations with a high production level have the highest oil consumption and create the highest pollution, with the potential for producing high levels of CO_2 per capita. Countries with lower productivities generate much lower

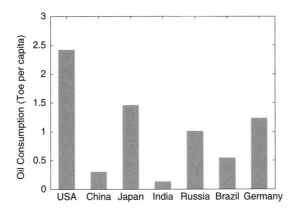

Figure 18.13 Oil consumption per capita of the seven biggest oil consumers in 2011, 1 toe = 41.87 gigajoules. (Data from Global Energy Statistical Yearbook 2012).

levels of CO_2. As these countries strive for a higher standard of living, their potential contribution to pollution will increase.

Figure 18.12 shows the highest ranking oil consumptions in 2011. Figure 18.13 gives the oil consumption per capita in these countries, which is proportional to CO_2 pollution and may be linked to climate change. Due to the high population in Asia, there is a huge demand for oil from Asia. The automobile sector is already growing very fast in both China and India, and the demand from these countries for energy increases rapidly.

19
Economics and entropy

I have not been everywhere, but it is on my list.
   Anon

Thus far we have made progress at the macroscopic level, without considering details of the microstructure of an economic system. In this sense, we have formulated physical macroeconomics. Our approach has introduced a new state function, the (economic) entropy. At the macroeconomic level, this can seem a somewhat mysterious quantity. A proper understanding of entropy is only possible by considering the system microstructure. Recalling eqn (1.4), we can say that entropy measures the total number of microstates a system may adopt that are consistent with the macrostate. As a result of evolution of the dynamical variables, the system will, over a sufficient time, visit all possible points in the system phase space.

Following the definition of Boltzmann entropy introduced in Section 1.4, we may say that in economics, the entropy is similarly a measure of the total number of available 'economic' states, whereas the energy measures the probability that any particular state in this 'economic' phase space will be realised. So, for example, in a system of traders it could be the total number of ways that the money and goods in the system can be distributed across the agents.

19.1 Entropy and probability

Basic concepts of probability theory have already been introduced in Chapter 3. Here we address the problem of distributing N different elements of a system into k different boxes or categories. Each box contains N_i elements, so we have $N = \sum_{i=1}^{k} N_i$. The number Ω of possibilities having N_1 elements in box 1, N_2 elements in box 2 etc. is given by

$$\Omega(\{n_i\}) = \frac{N!}{\prod_{i=1}^{k} N_i!}. \tag{19.1}$$

The *relative* number x_i of elements in a box/category is defined as

$$x_i = N_i/N, \tag{19.2}$$

resulting in $\sum_{i=1}^{k} x_i = 1$. For this reason, each x_i may also be interpreted as the probability that an element is in a box, category or state i.

The probability of a *particular distribution* $\{N_i\}$ (i.e. with N_1 specific elements in box 1, N_2 specific elements in box 2 etc.) may then be written as

$$p\{n_i\} = \Omega \prod_{i=1}^{k} \left(\frac{N_i}{N}\right)^{N_i} = \Omega \prod_{i=1}^{k} x_i^{N_i} = \Omega p_0, \tag{19.3}$$

where we have defined p_0 as

$$p_0 = \prod_{i=1}^{k} x_i^{N_i}. \tag{19.4}$$

If all boxes have the same size (equivalently: all categories are equally likely), i.e. $x_i = 1/k$, we have $p_0 = (1/k)^N$.

Example 19.1 A car dealer has six different cars for display in three show rooms. He puts three cars in the first show room, one in the second, and two in the third room. In how many ways may the cars be displayed?

We have $N = 6, k = 3, N_1 = 3, N_2 = 2, N_3 = 2$ and thus obtain $\Omega(3; 1; 2) = 6!/(3!1!2!) = 60$. There are 60 possibilities to display these six cars in the above manner.

The probability to find three *specific* cars in the first show room, one specific car in the second, and the other two cars in the third show room is obtained from eqn (19.3) with $p_0 = (3/6)^3(1/6)^1(2/6)^2 = 0.00231$. This results in $p \simeq 0.139 = 13.9\%$.

The statistical definition of (economic) entropy is given by

$$S_e = \ln \Omega, \tag{19.5}$$

where Ω is defined via eqn 19.1. This is equivalent to the definition of entropy in statistical physics, see eqn (1.4).[1]

For large numbers ($N \gg 1$), Stirling's approximation may be applied for the factorials in the expression for Ω,

$$\ln N! = N \ln N - N. \tag{19.6}$$

Thus the entropy may be written as

$$S_e = N \ln N - \sum n_i \ln n_i. \tag{19.7}$$

Introducing $x_i = n_i/N$ this reduces to

$$S_e = -N \sum x_i \ln x_i. \tag{19.8}$$

The additive property of entropy enters via the prefactor N, i.e. the total number of elements.

---

[1] From thermodynamics it is known that the empirically introduced entropy function of a system (A+B), combined of two subsystems A and B, is given by $S(A + B) = S(A) + S(B)$. The number of possible micro states of the combined system is given by $\Omega(A+B) = \Omega(A)\Omega(B)$. Writing $S = k_B \ln \Omega$ thus maintains the additive property of the entropy.

Exercise 19.1 The reader might wish to reproduce the above relationship which is one of the key relationships in statistical physics.

Note that we had introduced the equivalent relationship in statistical physics already in Chapter 1, as $S = -k_B \sum p_i \ln p_i$ (eqn (1.3)). In this case the entropy is that of a macroscopic system that finds itself in a state i with probability p_i. The prefactor k_B (Boltzmann constant) ensures that the dimension of entropy has the physical dimension of J/K, as required from thermodynamics.

19.2 Entropy as utility function

In Section 15.3 we mentioned that in *macroeconomics* the entropy function S_e takes the role of a production function of the macroscopic economic system, and we discussed this further in Sections 15.5 and 15.6. In *microeconomics* the entropy function may be used to find the optimal number of different professionals in a company, the best choice of different commodities in a household, the best choice of stocks or bonds in a portfolio. In neoclassical economics this function is called *utility function*.

In the following we shall discuss the application of entropy as utility function by restricting ourselves to the simplest case of binary systems with only two different categories, containing N_1 and N_2 elements, respectively. This may, for example, describe the case of a company employing only two types of workers, N_1 skilled and N_2 unskilled.

We shall first consider the case where the total number of elements, $N = N_1 + N_2$, is not fixed. This leads to the binary entropy/utility function

$$S_e(N_1, N_2) = N \ln N - N_1 \ln N_1 - N_2 \ln N_2, \tag{19.9}$$

which is plotted in Figure 19.1 as a function of N_1, at constant number N_2 (thus the total number $N = N_1 + N_2$ is *not* fixed).

It is seen that the utility function grows sub-linearly. Doubling the workforce will not double the utility of a company. While the first additional people may still contribute to the utility in a nearly linear way, this is less and less so for any further additional people.

Rewriting eqn (19.9) by setting $N_1/N = x$ and thus $N_2/N = 1 - x$ we obtain

$$S_e(x) = -N[x \ln(x) + (1 - x) \ln(1 - x)]. \tag{19.10}$$

Note that this is the expression for the *mixing entropy* which we had used earlier in our model of trade in a two-island world, eqn (17.1). Figure 19.2 shows a plot of this binary utility function per capita, $S_e(x)/N$, featuring a maximum at $x = 1/2$.

In neoclassical economic theory the following Cobb-Douglas utility function F_{CD} is used to describe many binary systems, such as companies with two different types of employees or markets with two different types of commodities

$$F_{CD} = N_1^\alpha N_2^{1-\alpha}. \tag{19.11}$$

The exponent α is called *elasticity* and is a free parameter, generally taken as $0.5 \leq \alpha \leq 0.7$ (Bofinger, 2003). This function is plotted in Figure 19.3. As is the case

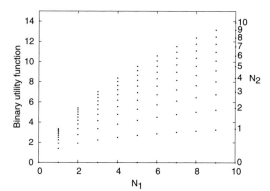

Figure 19.1 Plot of the binary utility function $S_e(N_1, N_2)$ of eqn (19.9) as a function of N_1 for different (fixed) values of N_2.

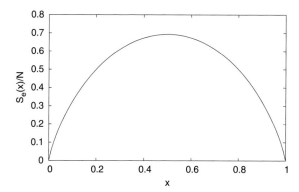

Figure 19.2 Plot of the binary utility function per *capita* S_e/N of eqn (19.10) as a function of x. This describes, for example, the production output for a company with a constant number N of staff at two different kinds of jobs. A fraction x might work, for example, in production, and a fraction $1 - x$ might work in sales. If there is no-one working in production ($x = 0$), nothing may be produced. If all workers produce ($x = 1$) there will be no sales and again there is no output. Without further constraints the maximum of utility is at equal numbers of production and sales, $x = 1/2$.

for the binary utility function S_e, F_{CD} increases sublinearly with N_1, however, when comparing it with Figure 19.1, we see that F_{CD} lies below S_e, thus leading to a lower output.

Again setting $N_1/N = x$ and $N_2/N = (1 - x)$ we obtain

$$F_{CD}(x) = Nx^\alpha(1 - x)^{1-\alpha}. \tag{19.12}$$

Both *relative* utility function $S_e(x)/N$ and the relative neoclassical Cobb-Douglas function $F_{CD}(x)/N$ are plotted in Figure 19.4 for comparison, using two different values of α. Again, we can see that our entropy based function S_e/N outperforms the Cobb-Douglas function.

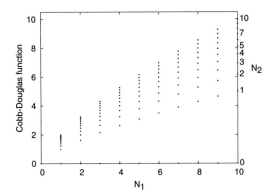

Figure 19.3 Plot of the neoclassical Cobb-Douglas utility function F_{CD} of eqn (19.11) as a function of N_1 in the range from 0 to 10 for values N_2 in the same range. Here the elasticity parameter has been set to $\alpha = 0.7$.

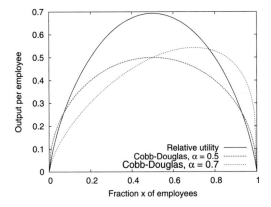

Figure 19.4 Plot showing relative utility function S_e/N of eqn (19.10) and the neoclassical Cobb-Douglas function $F_{CD}(x)/N$ of eqn (19.12) for $\alpha = 0.5$ and $\alpha = 0.7$. These functions might describe the output per employee for a company with N staff, working at two kinds of jobs, at a staff ratio of x and $1 - x$. The output as obtained using S_e is clearly superior.

The empirical Cobb-Douglas utility function contains the adjustable elasticity parameter α which allows for the adaptation of the function to data. In contrast, S_e emerges from an entropy-based theoretical framework containing no free parameter and thus has true predictive power.

19.3 Optimisation problems

Market traders have to decide the amount of a specific commodity on offer, and the corresponding price. Companies have to determine the number of people working in different positions, and the corresponding salaries. Salaries or prices lead to questions such as 'Should a market stall only offer expensive flowers or only cheap apples'?

In companies, the optimal number of people will depend on optimal productivity and wages for different jobs. Surely a company should not only hire directors with high income, or only low income workers, but which combination will be optimal for production?

In mathematical terms such optimisation problems are expressed as the maximisation or minimisation of a function, subject to constraints. In Section 11.2 we introduced the concept of a Lagrangian to fulfil such a task (see also Appendix 11.7). Here we will consider the problem of minimisation of cost C under the constraint of the given production or utility function S_e.

The Lagrangian $\mathcal{L}$, whose minimum we seek to determine, may then be written as

$$\mathcal{L}(\{N_i\}, m_L) = C(\{N_i\}) - m_L S_e(\{N_i\}), \tag{19.13}$$

where m_L is a yet undetermined Lagrange multiplier.

Before considering again our previous example of a binary system with a fixed total number of constituents, N, let us note that equivalent Lagrangian formulations feature in thermodynamics. In this case, $\mathcal{L}$ might be interpreted as Helmholtz free energy $F = U - TS$. Here, minimisation of F corresponds to the minimisation of internal energy U under the constraint of maximal entropy S. Temperature T acts as Lagrange multiplier. Similarly, in the presence of a volume V as a second constraint, one obtains the Gibbs free energy, $G = U + pV - TS$, with pressure p acting as a second Lagrange multiplier.

We will in the following consider the example of a company with N_1 permanent and N_2 temporary staff with wages per hour of C_1 and C_2, respectively. The total number of staff $N = N_1 + N_2$ is constant, and the budget restriction is given by the total cost C of the labour, $C = N_1 C_1 + N_2 C_2$. The Lagrangian is thus given by

$$\mathcal{L} = N_1 C_1 + N_2 C_2 - m_L \left[(N_1 + N_2) \ln(N_1 + N_2) - N_1 \ln N_1 - N_2 \ln N_2 \right]. \tag{19.14}$$

The stationarity conditions with respect to variation of N_1 and N_2 are

$$\partial \mathcal{L}/\partial N_1 = -m_L \left[\ln(N_1 + N_2) - \ln N_1 \right] + C_1 = 0,$$
$$\partial \mathcal{L}/\partial N_2 = -m_L \left[\ln(N_2 + N_2) - \ln N_2 \right] + C_2 = 0. \tag{19.15}$$

Solving for N_1 and N_2 at constant total number N and introducing the relative numbers of permanent and temporary staff, x and $1 - x$, we obtain

$$x = N_1/N = \exp(-C_1/m_L),$$
$$1 - x = N_2/N = \exp(-C_2/m_L). \tag{19.16}$$

The calculated distribution of people in the different jobs follows a Boltzmann distribution.

Solving for m_L results in

$$-1/m_L = \ln x/C_1 = \ln(1 - x)/C_2, \tag{19.17}$$

and thus

$$x^{C_2/C_1} + x = 1. \tag{19.18}$$

The relative numbers of permanent and temporary staff x and $1 - x$, the Lagrange parameter m_L, the utility or output per person S_e/N, and the wages per person C/N may now be straightforwardly calculated from the wages C_1 and C_2.

Returning again to eqn (19.16), we also see that the relative number of staff N_1 that are employed at a wage C_1 is given by a Boltzmann distribution. Such a result has already been found for the calculation of a demand function in our simple model of trade, eqn (17.7).

Example 19.2 Calculation of budget restrictions

A beer garden has N_1 permanent and N_2 temporary employees. The wages per hour are $C_1 = €15/h$ for the permanent and $C_2 = €7.5/h$ for the temporary, less-trained, staff, who will generally work less efficiently.

From eqn (19.18) we obtain $x = 0.382$, computed (generally) numerically, or graphically (see below), resulting in $m_L = 15.59$. The mean utility is obtained from eqn (19.10) as $S_e/N = 0.66$ and the mean wages as $C/N = 10.35$.

For a staff of $N = 10$ people, the optimal production of service is thus obtained for four permanent and six temporary employees. The total labour costs are €105/h (€103.50/ h) and the total output per hour is $S_e = 6.6$ (e.g. barrels of beer per hour). If the demand rises to $S_e = 10$ (barrels of beer per hour), 15 people should be hired at the same ratio.

Note that the application to actual problems requires the wages to correspond to the productivity at the job: professional people at wages $C_1 = €15/h$ are expected to be more productive than less well trained people at wages $C_2 = 7.50/h$, but there is generally *no linear* relationship between wages and productivity. In addition, the Lagrange multiplier m_L, which is the mean price of the system, must be scaled to the specific problem (for the example above it relates to the average amount of beer served and sold per waiter).

Example 19.3 Graphical solution

The optimal production at minimal costs may also be obtained from a graphical solution. Figure 19.5 shows three different functions between the relative number of temporary staff $x_2 = 1 - x$ and permanent staff, $x_1 = x$, intersecting at the ratio of optimal production, (x_1^o, x_2^o). These functions are

1. the relation between the two types of staff, $x_2 = 1 - x_1$.
2. the curve of maximal production, $x_2 = x_1^{C_2/C_1}$, as obtained from the Lagrange stationarity.
3. the condition due to the budget restriction, $C = N_1 C_1 + N_2 C_2$ results in $x_2 = C/(NC_2) - C_1/C_2 x_1$.
   The minimal budget, C/N, may be constructed by drawing a line starting from the point (x_1^o, x_2^o) of optimal production with slope C_1/C_2. This line intersects the y-axis at $C/(NC_2)$. The budget restriction is the tangent of the so-called iso-production line.
4. A possible fourth line is the iso-production line, which is used in neoclassical theory. It is obtained by solving for the production function $S(x_1, x_2)$ for x_2 at

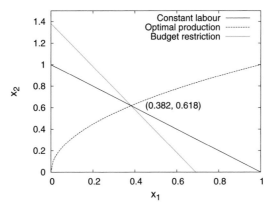

Figure 19.5 Graphical solution of optimal production at minimal costs: Three functions lead to the optimal point $(x_1, x_2) = (0.382, 0.618)$: the relative number of employees, the budget restriction, the line of optimal production.

constant S^o (iso-production). In calculus-based theory, the entropy function per person S_e/N is given by eqn (19.10). But this is an implicit function and cannot be solved analytically for x_2 at constant S^o. For this reason, the iso-production line is omitted in Figure 19.5.

The graphical solution leads to the same values of x_1 and x_2 as reported above, the maximal output, $S^0 = 0.66$ may only be determined numerically.

Neoclassical economics uses the Cobb-Douglas function F_{CD} of eqn (19.11) in the Lagrangian, eqn (19.13), to compute cost minimisation. We will in the following show that, for our example above, this results in higher mean wage cost at a lower mean output than what is achieved when using the entropy based utility function S_e.

Example 19.4 Computation of optimal production using the Cobb-Douglas function

The stationarity condition of the Lagrange function now gives

$$\partial \mathcal{L}_{CD}/\partial N_1 = C_1 - m_L \alpha N_1^{\alpha-1} N_2^{1-\alpha} = 0,$$
$$\partial \mathcal{L}_{CD}/\partial N_2 = C_2 - m_L (1-\alpha) N_1^{\alpha} N_2^{-\alpha} = 0. \tag{19.19}$$

The relative numbers of permanent, x_1, and temporary staff, x_2, at constant total number $N = N_1 + N_2$ are then obtained as

$$x_1 = \left(1 + \frac{1-\alpha}{\alpha} C_1/C_2 \right)^{-1},$$

$$x_2 = \left(1 + \frac{\alpha}{1-\alpha} C_2/C_1 \right)^{-1}. \tag{19.20}$$

Setting, as before, $C_1 = €15$ and $C_2 = €7.50$, and a value of $\alpha = 0.7$ (adhering to condition $0.5 \le \alpha \le 0.7$) results in $x_1 = 0.5385$ and $x_2 = 0.4615$. The mean output is

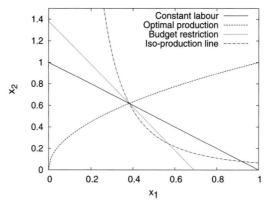

Figure 19.6 Graphical solution of optimal production at minimal costs using the Cobb-Douglas utility function. Four functions intersect at the optimal point $(x_1^o, x_2^o) = (0.538, 0.462)$. The functions relate to the relative number of employees, the budget restriction, the optimal production line, and the iso-production line. The solution depends on the choice of the elasticity parameter α.

given by $f_{CD}^0 = x_1^\alpha x_2^{1-\alpha} = 0.5141$ and the mean wage cost by $C/N = C_1 x_1 + C_2 x_2 = 11.5388$. This is inferior to our previous result of $S_e = 0.66$ and it can be shown that this is the case for all values of α. Also, the value for the Lagrange multiplier $m_L = 22.44$ exceeds that obtained from our entropy approach.

For completion we show (in Figure 19.6) a graphical solution. Four functions $x_2 = f(x_1)$ intersect in the point (x_1^o, x_2^o) of optimal production at minimal costs.

1. As before, the relation between the two types of staff is given by $x_2 = 1 - x_1$.
2. The curve of maximal production as obtained from the stationarity condition of the Lagrangian is given by $x_2 = \frac{1-\alpha}{\alpha} \frac{C_1}{C_2} x_1$.
3. As before, the condition due to the budget restriction, $C = N_1 C_1 + N_2 C_2$, results in $x_2 = C/(NC_2) - C_1/C_2 x_1$. The budget restriction is the tangent of the iso-production line.
4. The iso-production line is found by solving the Cobb-Douglas function for x_2 at constant production f_{CD}^o, $x_2 = f_{CD}^o{}^{\frac{1}{1-\alpha}}/x_1^{\frac{\alpha}{1-\alpha}}$.

The four lines meet at the point of optimal production at minimal costs, which depends on the choice of elasticity α, set to $\alpha = 0.7$ in Figure 19.6.

Again we see that the Cobb-Douglas function is not the optimal utility function in microeconomics. The existence of the free parameter α allows for a closer fit to experimental data, but at the same time this limits the predictive power of the neoclassical approach.

20

Approaches to non-equilibrium economics

The most ordinary things are to philosophy a source of insoluble puzzles. With infinite ingenuity it constructs a concept of space or time and then finds it absolutely impossible that there be objects in this space or that processes occur during this time.... the source of this kind of logic lies in excessive confidence in the so-called laws of thought.

Ludwig Boltzmann (1844–1906)

In Chapter 15 we developed an approach to economics by analogy to thermodynamics in which temperature is replaced by a parameter termed the dither. In an ideal gas, the temperature can be identified with the mean internal energy of the system; similarly, in an ideal economic system, the dither can be identified with the mean capital, or for nation state economies, the GDP per capita. Taking the ideas further we showed that two different subsystems, each at a different dither, could interact and, via a Carnot type process, generate new wealth and so give rise to economic growth. Over time, the dither associated with each subsystem can then clearly change.

In the first part of the book we saw how, in order to properly account for the statistics of financial fluctuations, a theory that assumed a constant volatility was a poor approximation and missed completely the fat tail risk associated with the log-price return fluctuations. So in economics we now find that a statistical mechanics in which the allowable states of the system are associated with a constant dither, leading to free energies which are proportional to the logarithm of a partition function, is not sufficient to describe fully economic systems. At the very least, a gradient in the dither in either time or space, that supports Carnot type processes, is needed.

To ask why such changes in the dither occur is a rather deep question. Ultimately one can expect it is associated with human behaviour and the response to economic situations. Different people can be expected to react in different ways, depending on their access to information, their ability to process information, and their general attitude to risk and so on. Furthermore any one person may not react in the same way at different times. Exploring this in detail is a part of behavioural economics and we shall not consider it further here.

The key point is that unlike physical systems where the temperature may be maintained by external heat baths, the dither in social systems is a self-generated phenomenon and to assume it is constant for an agent, whether the agent is a nation, a company, a community or an individual, may not be always be appropriate. To account for this, we suppose that the phenomenon can be captured by allowing the dither to fluctuate. Limiting the fluctuations to adiabatic processes we show how they can be incorporated with a generalized thermodynamics and statistical mechanics. We do this via the so-called *superstatistics*, developed by Beck and Cohen (2003) and Abe *et al.* (2007). This may offer a general framework for treating non-equilibrium stationary states of complex systems such as are found in physics and economics. We will also relate this to the *nonextensive statistical mechanics* approach by Tsallis (see for example Gell-Mann and Tsallis, 2004). The possibility of power law distributions will be of relevance also to our discussion of income and wealth distributions in Chapter 21.

20.1 Basic concepts

To begin, consider a complex system driven away from equilibrium by some kind of external force. A physicist might think of a particle diffusing within an inhomogeneous environment; an economist might think of complex networks of agents, such as cities or companies, where each strives to maintain its internal state in the competitive position by extracting energy or other resources from the environment. In the earlier chapters we have seen inhomogeneity in the volatility of a long financial time series. This volatility is a manifestation of the outcome of a group of agents engaged in trade.

The idea now is that such a system can be broken down into an ensemble of smaller sub-systems. Within each member of the ensemble it is assumed that the internal relaxation processes are sufficiently rapid, so that the system can adopt a local equilibrium state, characterized by a particular value of λ (or temperature for a physical system). Different members of the ensemble may thus operate at different values of λ. Alternatively, the value of λ might differ at different times, but the time scale over which such changes in λ are evident is much larger than the time characteristic of the internal relaxation processes. In previous chapters we have seen how for financial time series, the relaxation of the log-return autocorrelation function is typically a few minutes. So a member of our ensemble in this case could correspond to a slice of the time series typically greater than a few tens of minutes. Each ensemble member is now supposed at any instant to be in 'local equilibrium', characterized by a particular value of λ. Over the long-term, the stationary distribution of the 'superstatistical' system is then a superposition of Boltzmann factors with different values of λ that describe the local states of the individual members of the entire ensemble.

How does this development affect our thermodynamic theory? Indeed, is it still possible to formulate a meaningful thermodynamic approach? Many authors have sought to tackle this question. Tsallis has proposed the use of a new entropy function (often called 'Tsallis entropy') to address the matter (see for example Gell-Mann and Tsallis, 2004). The problem with this approach is that the entropy proposed appears to have no fundamental basis in physics or economics. A different route to the characterization of non-equilibrium systems goes back to Beck and Cohen (2003). It is rooted firmly in orthodox statistical mechanics and, as we shall see, follows logically from the theory we have developed thus far in this book.

20.2 A generalization of thermodynamics to non-equilibrium systems

Following the paper by Abe *et al.* (2007), we begin by considering a system that admits fluctuations of two independent random variables, ϵ (energy) and β (proportional to an inverse temperature), which can take on values (events, microstates) ϵ_i and β_j. We denote the joint probability function for the event (ϵ_i, β_j) as $p_{ij}(\epsilon, \beta)$. The Boltzmann-Gibbs entropy for such a system is then given by

$$S[E, B] = -\sum_{i,j} p_{i,j}(E, B) \ln p_{i,j}(E, B). \tag{20.1}$$

In writing this expression, E and B are dummy variables and will be used to differentiate additional entropies, to be introduced below.

Now invoke Bayes' rule (see Section 3.3) that allows us to relate the joint probability to the conditional probability, viz:

$$p_{ij}(E, B) = p_{ij}(E|B)p_j(B) = p_{ij}(B|E)p_i(E). \tag{20.2}$$

Using this expression we can re-express eqn (20.1) as

$$S[E, B] = S[E|B] + S[B], \tag{20.3}$$

where the *conditional entropy* $S[E|B]$ is given by

$$S[E|B] = \sum_j S[E, \beta_j)p_j(B), \tag{20.4}$$

with

$$S[E|\beta_j) = -\sum_i p_{ij}(E, B) \ln p_{ij}(E, B). \tag{20.5}$$

Since we have summed over the ϵ_i, $S[E|\beta_j)$ is a function of the β_j only.

Replacing the sums over j by integrals over β and introducing a global probability density $f(\beta)$ for the distribution of the values of β we obtain

$$S[E, B] = S[E|B] + S[B] = \int d\beta f(\beta) S[E|\beta) - \int d\beta f(\beta) \ln f(\beta). \tag{20.6}$$

Recall now that the probability distribution for a system in equilibrium is given by a canonical (Boltzmann) distribution function $p(\epsilon_i) = \exp(-\beta\epsilon_i)/Z(\beta)$ where Z is a normalization function, usually called the partition function. It is related to a Helmholtz 'free' energy $F(U, T) = U - TS = -\beta^{-1} \ln Z$. This canonical distribution function follows when we maximize the Boltzmann entropy subject to constant energy $U = \sum_i \epsilon_i p(\epsilon_i)$ and β emerges as a Lagrange multiplier.

Bearing in mind the above discussion, we now identify the conditional probability distribution function, $p(\epsilon_i|\beta)$, with this Boltzmann distribution, i.e. $p(\epsilon_i|\beta) = \exp(-\beta\epsilon_i)/Z(\beta)$, but where β is in what follows now assumed to be a second random variable, rather than a constant. Introducing this function into eqn (20.6) yields the basic result

$$S[E, B] = \overline{\beta U(\beta)} + \overline{\ln Z(\beta)} + S[B], \tag{20.7}$$

where the average of an arbitrary observable Q over the fluctuating variable β is defined as $\overline{Q(\beta)} = \int d\beta f(\beta) Q(\beta)$ and $U(\beta) = \sum_i \epsilon_i p(\epsilon_i|\beta)$.

This is the generalization of the laws of thermodynamics to non-equilibrium systems under the assumptions outlined above. To check that our expression reduces to well known thermodynamics, we may choose $f(\beta) = \delta(\beta - \beta_0)$, i.e. for the case of non-fluctuating β. It follows that $S[B] = 0$ (a result that is best obtained from a limit of the discrete version of the entropy expression) and we obtain

$$S[E, \beta_0] = \beta_0 U(\beta_0) + \ln Z(\beta_0). \tag{20.8}$$

This is indeed the familiar relation for the Helmholtz free energy in equilibrium thermodynamics and our zeroth model of economics. However, what happens when we have a more general expression for the probability density function, $f(\beta)$? Indeed is there any systematic way to deduce the distribution function $f(\beta)$?

20.3   A generalized Boltzmann factor

Recognizing that the state conditional on β taking a specific value is in equilibrium, we may express the unconditional probability $p(\epsilon)$ as follows:

$$p(\epsilon) = \int f(\beta) p(\epsilon|\beta) d\beta = \int f(\beta) \exp(-\beta\epsilon)/Z(\beta) d\beta. \tag{20.9}$$

We may now introduce the new function

$$\tilde{f}(\beta) = \frac{\beta}{c} \frac{f(\beta)}{Z(\beta)}, \tag{20.10}$$

which is another normalized probability distribution, where c is the corresponding normalization constant. This allows the expression involving the energy function in equation 20.7 to be rewritten as

$$\overline{\beta U(\beta)} = c \sum_i \epsilon_i B(\epsilon_i). \tag{20.11}$$

Here $B(\epsilon_i)$ is called the *generalized Boltzmann factor* for a non-equilibrium system (Beck and Cohen, 2003), given by

$$B(\epsilon_i) = \int \tilde{f}(\beta) \exp(-\beta\epsilon_i) d\beta, \tag{20.12}$$

i.e. $B(\epsilon_i)$ is the Laplace transform of the new distribution function $\tilde{f}(\beta)$. (The expression *superstatistics* refers to $B(\epsilon_i)$ representing the 'statistics of the statistics' $\exp(-\beta\epsilon_i)$ of the different subsystems making up the complex system under consideration.)

One distribution function $\tilde{f}(\beta)$ that has been used by a variety of authors (for example Touchette and Beck, 2005) is the Gamma distribution,

$$\tilde{f}(\beta) = \frac{1}{b\Gamma(d)} \left(\frac{\beta}{b}\right)^{d-1} \exp(-\beta/b), \tag{20.13}$$

where $\Gamma(z)$ is the Gamma function and b, d are positive parameters.

Using this expression, we can now perform the integration over β in the expression for the generalized Boltzmann factor to obtain

$$B(\epsilon_i) = \int_0^\infty \exp(-\beta\epsilon_i)\tilde{f}(\beta)d\beta = (1 + b\epsilon_i)^{-d}. \tag{20.14}$$

This function indeed exhibits fat tails providing $0 < d < 1$ and the alert reader will notice immediately that it is exactly the distribution we used earlier in Chapter 10 to characterize high frequency stock fluctuations [1] with exponent $\alpha = d$ (see eqn (10.28)). Clearly this function is a potential candidate for the description of fluctuations in finance and economics.

A particular property of the function defined by eqn (20.13) is that it is sharply peaked about the value $\beta_0 = bd$. For such a function, Abe *et al.* (2007) showed that the approach yields a model of the zeroth order type with a slightly different energy and modified averages. The derivation is straightforward.

If $\tilde{f}(\beta)$ is sharply peaked, then it is possible to compute the integral on the RHS of eqn (20.12) as a power series in the moments (Beck and Cohen, 2003), to obtain

$$B(\epsilon_i) = e^{-\beta_0\epsilon_i}\left(1 + \frac{1}{2}\sigma^2\epsilon_i^2 + \cdots\right). \tag{20.15}$$

The variance σ^2 is given by

$$\sigma^2 = \int d\beta\beta^2\tilde{f}(\beta) - \left(\int d\beta\beta\tilde{f}(\beta)\right)^2. \tag{20.16}$$

Using eqns (20.15) and (20.16) we can evaluate eqn (20.11) to obtain

$$\overline{\beta U(\beta)} = c\langle E\rangle + \frac{c}{2}\sigma^2\langle E^3\rangle + \cdots, \tag{20.17}$$

where the brackets $\langle\ldots\rangle$ denote an un-normalized canonical average for a system at the temperature β_0, i.e. $\langle E^m\rangle = \sum_i \epsilon_i{}^m e^{-\beta_0\epsilon_i}$.

The average energy, however, is multiplied by neither β nor β_0. Rather it is multiplied by c, which since the function f is sharply peaked, is close to β_0. Also, there are higher order correction terms, with leading order E^3.

So even when we have fluctuations of the dither (or β) that take the system out of equilibrium, we still seem to be able to exploit a thermodynamics type of description. We use the entropy given by eqns (20.3) to (20.5), which for the particular function $f(\beta)$ defined by eqn (20.13) leads to the generalized Boltzmann function of eqn (20.14) with power law tails. We also have the renormalized inverse temperature, c, which is in any event close to β_0, and the internal energy, U, which now is no longer linear, but is a power series of the un-normalized energy $\langle E\rangle$.

There are other options one may chose for the function $f(\beta)$. Not all, however, will necessarily lead to fat tails. Choosing β^{-1} (i.e. what we called 'dither') rather than β as the fluctuating parameter and lognormal functions are both considered by Abe. The

---

[1] As in Section 7.3, here we identify the log-return with a velocity, ϵ is the kinetic energy, $\propto v^2$.

former leads to a marginal distribution function, eqn (20.9), with exponential decays in $\epsilon^{1/2}$.

Lognormal fluctuations have been shown to be relevant for turbulent flows in fluids. Since financial data and income data both follow power-law distribution functions, we might be inclined to think that all economic fluctuations behave in this way. However, we cannot be sure of this and to distinguish between all the many possible options requires comparison with data. More detailed empirical studies of economic systems would greatly help this task.

20.4 Superstatistics and non-extensive statistical mechanics

A different approach for dealing with non-equilibrium systems was taken by Tsallis (Tsallis, 1988, Gell-Mann and Tsallis, 2004). In this *non-extensive statistical mechanics*, (Tsallis-)entropy is defined as

$$S_q = \frac{1}{q-1}\left(1 - \sum_i p_i^q\right), \tag{20.18}$$

resulting in a generalized Boltzmann factor of the form

$$B_q = (1 + (q-1)\beta_0\epsilon_i)^{-1/(q-1)}, \tag{20.19}$$

where q is called the *entropic index*. The (equilibrium) expressions for Boltzmann-Gibbs entropy and the Boltzmann factor are recovered for $q = 1$. We note that for $d = 1/(q-1)$, B_q is identical to the generalized Boltzmann factor that we obtained from superstatistics, when using the Gamma function for modelling temperature fluctuations, see eqn (20.14).

Expanding the generalized Boltzmann factor B_q defined by eqn (20.19) in a Taylor series yields

$$B_q = e^{-\beta_0\epsilon_i}\left(1 + \frac{1}{2}(q-1)\beta_0^2\epsilon_i^2 - \frac{1}{3}(q-1)^2\beta_0^3\epsilon_i^3 + \cdots\right). \tag{20.20}$$

Comparing this with the expansion for B in eqn (20.15), which is universal for all distributions in the limit of *small* values of ϵ_i (Beck and Cohen, 2003), we can immediately identify $q = \frac{\langle \beta^2 \rangle}{\langle \beta \rangle^2}$. Tsallis statistics might thus be regarded as a special case of superstatistics.

20.5 Superstatistics for an ideal system

In Section 16.1 we considered a simple model of an economic system, based on an analogy with the ideal gas where the microstructural subunits are independent. Abe *et al.* (2007) also consider their concept of superstatistics in this context and it is relevant and interesting to present their analysis.

To proceed we must first address whether it is possible to determine the optimum distribution of fluctuations of the dither in this superstatistical system of N non-interacting subunits. If we have no prior information, it may be obtained by maximizing the entropy, subject to the constraint that the function $f(\beta)$ is normalized,

$$\frac{\delta}{\delta f}\left[S[E, B] - \alpha\left(\int d\beta f(\beta) - 1\right)\right] = 0, \tag{20.21}$$

where α is a Lagrange multiplier. Since the short time dynamics are averaged, the entropy $S[E, B]$ will be a functional only of $f(\beta)$. Recalling the definition of $S[E, B]$, we obtain immediately the solution

$$f(\beta) = \frac{\exp S[E|\beta]}{N}, \tag{20.22}$$

where the entropy is given by $S[E|\beta] = \beta U(\beta) + \ln Z(\beta)$.

Now we return to the details of our system of independent subunits. Recall that the conditional probability of the energy taking a particular value is given by $p(\epsilon_i|\beta) = \exp(-\beta\epsilon_i)$ with the partition function $Z(\beta) = \beta^{-N}$. The local internal energy of the system is $U = N/\beta$. We can now compute the conditional entropy as

$$S[E|\beta] = -N\ln\beta + \text{terms independent of } \beta. \tag{20.23}$$

As a result we see that the function $f(\beta)$ given by eqn (20.22) is

$$f(\beta) \propto \beta^{-N}. \tag{20.24}$$

This is a power law distribution with no peaks outside the origin and it is normalizable only over a finite range of β. Abe *et al.* (2007) argue that for physical systems, it is reasonable to expect the function to be so normalizable. Here, we assume that a similar situation should be expected for economic systems, in which case it is possible to compute

$$\tilde{f}(\beta) = \frac{\beta f(\beta)}{cZ(\beta)} \propto \beta. \tag{20.25}$$

In a separate paper, Touchette and Beck (2005) show that if for small β the distribution $\tilde{f}(\beta)$ scales as $\tilde{f}(\beta) \propto \beta^\gamma$, and $\gamma > 0$, then the generalized Boltzmann factor, eqn (20.12), decays for large values of the energy as $B(\epsilon_i) \propto \epsilon_i^{-1-\gamma}$.

For the ideal system we have $\gamma = 1$. Comparing this result with eqn (20.19) it follows that $1/(q - 1) = 2$, and thus $q = 1.5$, similar to the values for the entropic index q obtained for various complex systems (Abe *et al.* 2007). Our value of $\alpha \simeq 1.84$, obtained empirically for minute-by-minute stock data as described in Section 10.4, corresponds to $q = \alpha^{-1} + 1 = 1.54$.

As was pointed out by Abe and colleagues, superstatistics is a very powerful approach to gaining insight into non-equilibrium systems. The use of conditional entropies offers a way to embed multi-scale superstatistical phenomena and obtain corresponding thermodynamics. The ideas are, at this time, relatively new and it seems certain that more remains to be uncovered using the method not just in physics, but also in economics.

21

The distribution of wealth in society

If you know how rich you are, you are not rich. But me, I am not aware of the extent of my wealth. That's how rich we are.

Imelda Marcos, former First Lady of the Philippines.

That income and wealth is not distributed uniformly in society appears obvious. But how are these attributes distributed? What is the form of the distribution density function $p(m)$, and does it depend on time, history or locations? These questions had been posed ever since Vilfredo Pareto over 100 years ago noticed that the rich end of the wealth distribution followed a power law, i.e.

$$p(m) \propto m^{-(1+\alpha)}, \quad \text{for large values of wealth } m, \tag{21.1}$$

and that this feature seemed to be universal (Pareto, 1897). Figure 21.1 shows as an example the cumulative distribution of wealth for the top richest people in the world in 2013, according to data taken from Forbes magazine. Many such distributions have been investigated, and in Figure 21.2 we show the spread of values for the *Pareto exponent* α as obtained for distributions of wealth from various sources.

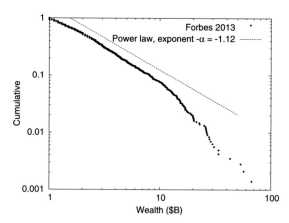

Figure 21.1 For large values of wealth m (shown in billion dollars), the cumulative wealth distribution function $C_>(m) = \int_m^\infty p(m)dm$ varies in the form of a power law, $C_>(m) \propto m^{-\alpha}$ (see eqn 21.1). Data shown is extracted from the 2013 Forbes list of 'super-rich' billionaires and is described by a power law with exponent $\alpha \simeq 1.12$ (Data: http://www.forbes.com/billionaires).

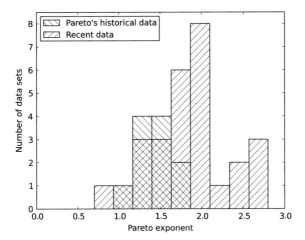

Figure 21.2 The distribution of Pareto exponents for mainly end of the twentieth century data appears to have widened and shifted to larger values, compared with an extract of Pareto's original nineteenth and sixteenth century data. The smallest values of the exponent α in the twentieth century data sets are associated with lists of the 'super-rich', as opposed to the top earners in income data sets. (For details of the sources of the various data sets see Richmond *et al.* (2006).)

To Pareto, and indeed to most physicists, the existence of such a power-law suggests that some fundamental dynamics is in play (see for example Chatterjee *et al.* 2005, or the review article by Richmond *et al.* 2006). Pareto himself proposed that people, in the course of their life, could move through the distribution in both directions, and this idea that a static distribution does indeed not imply a static society has formed the basis for the recent studies by a number of physicists.

21.1 Gibrat's model

A simple model for wealth distribution, formulated by Robert Gibrat in 1931, is based on the so-called *multiplicative stochastic process*, where the value of a variable at the discrete time variable $t + 1$, in our case wealth m_{t+1}, is proportional to its value at time t, i.e.

$$m_{t+1} = c_t m_t. \tag{21.2}$$

Gibrat's terminology of 'proportionate effect' refers to the proportional change of a quantity being independent of the quantity itself, i.e. $(m_{t+1} - m_t)/m_t = c_t - 1$. From eqn (21.2) we obtain

$$m_{t+1} = \Pi_{i=0}^{t} c_i m_0, \tag{21.3}$$

where the c_i are time-dependent constants and $m_0 = m(t = 0)$. Taking the logarithm of both sides and applying the central limit theorem (see Section 3.2), we immediately find that $\ln m_t$ is normally distributed, i.e. m_t is distributed according to a log-normal distribution (see eqn (6.17)).

Gibrat's model has been used to describe empirical data for the lower end of the wealth distribution. By introducing a threshold of minimum income, as in the model of Champernowne from 1953, it is also possible to obtain Pareto's power law form for large wealth.

21.2 The generalized Lotka-Volterra model

Robert Gibrat died in 1980, yet after his publication in 1931 little more was done to develop his approach to wealth and income distributions and the power law of Pareto remained more of a curiosity than being confirmed by theory. This may reflect the increased interest in macro-economics. It perhaps also reflects the fact that log-normal distributions and the associated Gaussians, as were also used in finance, were easier to deal with.

Only Mandelbrot was continuing to speak of the importance of power law distributions and offering some ideas as to their origin (see Section 21.3), but it was not easy to see how his ideas fitted into the orthodoxy. In statistical physics this changed with new insights into the idea of scaling and their implications in statistical physics by Wilson in 1974, as we already mentioned in Section 12.3.

However, such thinking did not permeate economics when it was dominated by powerful groups that favoured existing approaches. Moreover, using the available data at the time, it was always easy to argue that the log-normal function provided just as good a fit as that of a power law. Given the permeation of scaling ideas and the development of ways to explore power laws in distribution functions, it is perhaps not surprising then that physicists sought to re-examine some of the power laws in economics, following their foray into finance.

So how may one devise a simple model that captures both the low and the high end of wealth, including the power law tails for which there was now more evidence (see for example the data of Figure 21.1, and also Yakovenko and Silva (2005))? This was addressed by Solomon and Richmond and colleagues in a series of papers at the beginning of the last decade (e.g. Richmond and Solomon, 2001, Solomon and Richmond, 2002, Malcai *et al.* 2002).

The model of Gibrat simply assumes that at each time step, the wealth associated with an agent changes in a random manner according to the character assumed for the stochastic process, c_t. The key idea of Richmond and Solomon was to introduce an additional process which redistributes a small fraction, $a > 0$, of an agent's wealth to other agents in the economic system.

The model is based on the so-called Generalized Lotka-Volterra model (GLV), and is described by the following equation,

$$w_i(t + \tau) - w_i(t) = (r_i(t) - 1)w_i(t) + aw(t) - c(w, t)w_i(t), \qquad (21.4)$$

where the time dependent function $w_i(t)$ is identified with the wealth of an individual agent i at time t. The $r_i(t)$ are random numbers picked from the same probability distribution (independent of i), with variance D. They represent the random part of returns to wealth during time t and $t + \delta$. The coefficient a represents, in a simplified way, wealth received by agents in terms of benefits, and services from their community, which is why it is proportional to the *average wealth* w of the community.

The coefficient $c(w, t)$ represents, again in a simplified way, external limiting factors, such as the finiteness of resources, technological innovations, and internal market effects, including competition. It thus parameterizes the general state of the economy. When c is large and positive, the economy may be thought to be growing; when it is negative, the economy is contracting. This particular term may also be thought of as an expression of inflation. If one identifies w_i with the real wealth, as opposed to its numerical value, then an increase in the numerical value of the total wealth measured by w, means that an agent with individual wealth w_i will lose, due to inflation, an amount proportional to both $w(t)$ and $w_i(t)$, i.e. $c(w,t) - cw(t)$ where c is a constant, such that the third term in eqn (21.4) becomes $-cw(t)w_i(t)$.

Taking the average of eqn (21.4) over all agents i gives an equation for the total wealth, w,

$$w(t + \delta) = (r(t) + a)w(t) - c(w,t)w(t). \qquad (21.5)$$

The complete derivation is given in Solomon and Richmond (2001).

Exercise 21.1 The reader is invited to show that in the continuum limit, when $\tau \to 0$, the *relative wealth* of an agent $x_i(t) = w_i(t)/w(t)$ satisfies the differential stochastic equation

$$\frac{dx_i}{dt} = \epsilon_i(t)x_i - ax_i + a, \qquad (21.6)$$

where the random numbers $\epsilon_i(t) = r_i(t) - r(t)$ are such that $\langle \epsilon_i(t) \rangle = 0$ and $D = \langle \epsilon_i(t)^2 \rangle$. The interesting aspect of this equation is that the relative wealth depends only on D and a. The details of the interactions $c(w,t)$ or the average growth, r, are absent.

Equation (21.6) has the form of the generalized Langevin equation (eqn (7.22)), where

$$F(x,t) = a(1 - x_i) \text{ and } G(x,t) = x_i. \qquad (21.7)$$

It follows that the (normalized) probability distribution function for the relative wealth $x_i(t)$ is

$$P(x) = \frac{(\alpha - 1)^\alpha}{\Gamma(\alpha)} \frac{\exp(-\frac{\alpha-1}{x})}{x^{1+\alpha}}, \qquad (21.8)$$

where $\alpha = 1 + 2a/D$. Interestingly we see that if a/D is constant, then even for large variations of $c(w,t)$ and average wealth $w(t)$, the dynamics of the relative wealth are time independent. Thus even for a non-stationary system, the relative wealth distribution function converges to a time invariant form which exhibits a Pareto-like fat tail. Other interpretations can be placed on the model, relating to stock market fluctuations and we refer the reader to the original paper for a discussion of these.

Richmond *et al.* (2006) showed, using income as a proxy for wealth, how the above distribution function describes UK income data over the period 1995 to 2005. Figure 21.3 features a fit of the cumulative distribution function obtained for eqn (21.8) to the data for the year 1995. Additional 1995 data, extending to weekly income up to £10000, featured the same power law decay (Richmond *et al.*, 2006).

21.3 Collision models

A physicist may interpret the underlying character of the Generalized Lotka-Volterra model in terms of agent collisions in the sense that one agent undergoes an exchange

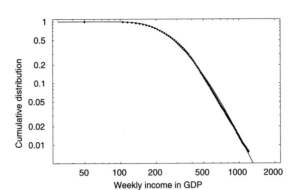

Figure 21.3 Least-squares fit of the Generalized Lotka-Volterra *cumulative* distribution function, $C_<(x) = 1 - \Gamma(\alpha, \frac{\alpha-1}{x})/\Gamma(\alpha)$, to the UK New Earnings Survey data 1995 for weekly incomes of employees in the UK (http://www.statistics.gov.uk). There is a noticeable deviation from the data for the income between 500 and 900 GBP in this two-parameter fit. (Note that one of the fit parameters is required for the scaling of the abscissa).

of income with each and every other agent in the system during a time step. The idea of modelling income distributions using this concept was made by Mandelbrot in 1960.

Mandelbrot's idea was that transactions between agents are similar to the pairwise exchanges in energy and momentum between molecules during collisions and so it should be possible to exploit all the methods of statistical physics developed for such processes. In 1986, the sociologist John Angle published the first in a series of papers that presented a simple stochastic agent model that essentially exploited Mandelbrot's proposal and which led to the emergence of inequality in wealth distributions. Angle's approach was based upon evidence attributed to archaeological excavations that suggested to him that inequality within a community first emerged as the agricultural revolution took hold and there grew an abundance of food. Part of the surplus produced could be taken away in an interaction ('collision') of agents by means of theft, taxation, extortion, voluntary exchange or gift.

The work was taken up and extended by Chakraborti and others in a string of papers (see Chatterjee *et al.* 2005). The resulting collision rule can be described simply: at an encounter, an agent i may save or set aside a fixed fraction λ_i of his/her assets, m_i. The remainder is put on the table, together with the assets of the second agent. A coin is then thrown to determine who takes a previously specified fraction ε of the total assets on the table leaving the remainder for the loser of the game. The governing equations for interacting agents i and j are thus given by

$$\begin{pmatrix} m_{i,t+1} \\ m_{j,t+1} \end{pmatrix} = \begin{pmatrix} \lambda_i + \varepsilon(1 - \lambda_i) & \varepsilon(1 - \lambda_j) \\ (1 - \varepsilon)(1 - \lambda_i) & \lambda_j + (1 - \varepsilon)(1 - \lambda_j) \end{pmatrix} \begin{pmatrix} m_{i,t} \\ m_{j,t} \end{pmatrix}, \tag{21.9}$$

and are iterated numerically in a Monte Carlo simulation. During this process, the net total value of assets possessed by the two agents is conserved. No new money is created during the interchange.

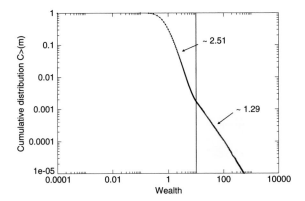

Figure 21.4 Computer simulation showing a cumulative wealth distribution with two pow-er-law regimes (Coelho *et al.* 2008b). The simulation uses a modification of the Slanina model and accounts for the observation that the Pareto exponents for data from super-rich lists are generally lower than the one for data from income distributions. The x-axis displays wealth divided by the mean wealth. (Figure reprinted from Coelho *et al.* (2008b), with permission from Elsevier.)

Even for the simplest case of uniform saving fractions, i.e. $\lambda_i = \lambda_j = const.$, this approach yields a steady state distribution for the m_i that can be fitted to the middle and lower range of empirical income distribution data. In order to yield a power law tail in the high end of the distribution function, the λ_i and λ_j need to be selected from a random distribution at each pairwise interaction. However, as shown by Repetowicz *et al.* (2005), the Pareto exponent α is exactly one. This approach thus does not reproduce the spread of values of α found empirically, see Figure 21.2.

In 2004, Slanina proposed a different exchange law whereby money is created in the pair-wise exchange process as a result of investment (Slanina 2004). The specific exchange rule proposed by Slanina is deterministic. At each encounter, agents exchange a constant fraction β of their individual assets. The joint value of assets of the agents is now not conserved during a pairwise encounter, rather it increases by a fraction $\varepsilon'(m_{i,t} + m_{j,t})$, where $\epsilon' > 0$ is a constant. This increase in wealth might correspond to exploitation of natural resources; according to Slanina it is ultimately due to energy flowing from the sun to the earth.

Interestingly, the functional form of the resulting distribution function of Slanina's model is equivalent to that obtained from the generalized Lotka-Volterra model dis-cussed above. The Pareto power law tail has an index $\alpha \approx 1 + 2\beta/\varepsilon'^2$ for large values of income. By making the exchange fraction β dependent on the income, it is also possible to achieve a double power law, as shown in Figure 21.4. In this way, one can account for the difference in exponents that is found in data for people qualified as 'super-rich' (for example in Forbes magazine, see Figure 21.1) and exponents obtained from income data.

An exchange rule similar to that proposed by Slanina may be derived from our discussion of economical Carnot cycles in Chapter 18. Imagine our global economic

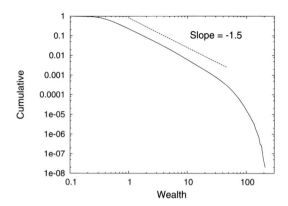

Figure 21.5 Computed wealth distribution for the agent model based on mini-Carnot cycles, eqn (21.10). The data may be approximated by a power law with exponent $\alpha = 1.5$.

system to consist of an ensemble of mini-Carnot cycles via which individuals constantly trade, moving from one mini-cycle to another. Suppose this is done in a random manner, then we might simulate this process at the microscopic level by a pairwise transfer matrix, obtained by rewriting eqn (18.5) for discrete time steps,

$$\begin{pmatrix} m_{i,t+1} \\ m_{j,t+1} \end{pmatrix} = \begin{pmatrix} 1-p & p \\ -(1-p) & (2-p) \end{pmatrix} \begin{pmatrix} m_{i,t} \\ m_{j,t} \end{pmatrix}, \tag{21.10}$$

for $m_j > m_i$. As with the other agent models described above, the long time distribution function of money across the many agents in the ensemble may be computed numerically. The result of such a computation, rescaled by the mean wealth, is shown in Figure 21.5; again the outcome is a power-law, but this time with an exponent of about $\alpha \simeq 1.5$.

21.4 Résumé

On the basis of the above results, there seem to be two essential ingredients if a Pareto tail in the distribution of wealth is to be obtained. First there must be a mechanism for redistribution of assets; second there must, during the encounter, be a creation or perhaps destruction of the total wealth of the agents. Non-conservation of assets during encounters seems to be important (else the value of the Pareto exponent is fixed to $\alpha = 1$, as discussed above). The system is not in the usual equilibrium state where a Gamma or Maxwell-Boltzmann distribution (see eqns (20.13) and (5.14), respectively) is the inevitable outcome.

Understanding the origin of income and wealth distributions has not been at the forefront of economic research for many years, ever since the work of Gibrat, and some economists have been critical of the approaches rooted in ideas that can be traced back to statistical physics. Others have been more positive. Thomas Lux has advocated the development of models more in line with standard principles of economic modelling (see Chatterjee *et al.* 2005). This would no doubt lead to a more detailed quantification of the interactions that occur within communities, as wealth is exchanged between individuals.

From the perspective of a physicist this seems equivalent to obtaining the detail of the interatomic potential. However, what is known from this latter work is that when considering the behaviour of large numbers of atoms, very few details of the interatomic potential are required in order to understand the existence of the Maxwell-Boltzmann distribution. That there are numerous molecular collisions and that the number of atoms within the system is large are the crucial attributes. It is these aspects that are necessary for the existence of critical points and phase transitions. Fine points of an inter-atomic potential determine details such as the precise values for the temperature of boiling and freezing of a liquid rather than the emergence of the Maxwell distribution of velocities per se. One is tempted to assume that for assemblies of agents, the fine details of the interactions are similarly not essential.

22
Conclusions and outlook

Ar scáth a chéile a mhaireann na daoine.
It is in each others' shadow that people live.
   Ancient Irish proverb

Understanding finance and economics has never been so important. The financial crisis of 2008 and events since then have given the issue renewed impetus. We began this journey arguing that physicists do have a contribution to make to the topic and we hope that having read this far you agree this to be the case. Unpacking the fluctuations in financial asset prices and introducing into economics the complementary concept of entropy, which allows quantification of the amount of randomness in the system, would seem to facilitate new insights into economic and financial systems. It is not, however, the end of the story. Rather it is merely the end of the beginning. We have learned that the fluctuations in material systems ultimately arise from the constant motion and collision of the many molecules that form the material. In the same way, the fluctuations in social and economic systems can arise from the many exchange processes between the agents in the system. As a result, objects and money become distributed across the agents and we see that the traditional idea of a representative agent, Quetelet's *l'homme moyen*, is no longer a sensible concept. Complex systems are heterogeneous and the key to understanding them is the use of a proper method of ensemble averaging.

There are numerous other aspects that we have not the space to cover in what is an introductory text. So, for example, in our analysis we limited ourselves to the use of one-minute data, however tic data that records every individual transaction is now also available. This opens up the option of looking at the waiting time between transactions (see for example Sabatelli *et al.* (2002)). Other workers have explored bid-ask spreads to see if more insight can be obtained. More significantly, entire order books are being made available to some researchers who can examine in great detail the impact of orders—both large and small—on the market prices. This opens up new and important routes to understanding and quantifying risk. A compendium of much recent work devoted to searching for regularities in financial data can be found in Takayasu *et al.* (2010).

Of more fundamental interest to researchers of social and economic systems is the nature and effect of the interactions between the underlying agents. That atoms interact via forces such as those mediated by electrical activity on the atoms and quantum mechanical effects is well known. Furthermore, simply knowing the broad features of atomic and molecular interactions can be sufficient to account for the fact that ensembles of large numbers of atoms can undergo phase transitions such as those from gas to liquid or metal to insulator, magnetic to non-magnetic phases. Of course, the precise temperature at which such phase changes occur, for example why water boils at 100°C, whereas oxygen boils at -183°C, depends very much on the details of the interaction. But the key to understanding phase change is to recognize the existence of interatomic interactions and use a proper statistical averaging of the many states of the atoms in the system, rather than worry unduly about obtaining a detailed understanding of the atom-atom interaction. In this sense, the physical theory is robust. Is it possible that a similar situation is true for economics?

We have already seen in Sections 11.4 and 11.5 that correlations between different assets exist and manifest themselves via the structure of minimal spanning trees and random matrices. Ultimately, the source of the correlations is activity of underlying agents who are trading the assets. At the root of all this are the basic human instincts of fear, greed, moral hazard (the ability to place the downside risk and associated losses with a third party), and other psychological traits that have determined risk taking since ancient times.

The recent financial crisis can now, with the benefit of hindsight, be seen to have roots in the actions of key agents within the banks acting incompetently. Board members ignored their obligations to the shareholders and owners. Managers encouraged compensation schemes to distort risk taking, such that traders took short-term profits at the expense of incurring longer term risk for the firm and its shareholders. Managers and decision makers then walked away with large compensation packages as these risky positions unravelled, leaving the firm with vast losses. As we now know, governments took over the losses, passing them on to taxpayers. As events unfolded over the first decade of the twenty-first century, senior politicians and regulators simply dismissed the unfolding stock bubble as irrational exuberance. It was said that we were evolving a new economy and this time things are different. The few people who recalled similar comments made during previous stock bubbles, such as those of 1927–9 and 1997–2001, were ignored. Sornette and Woodard (2010) have published an interesting indictment of the attitudes of bankers and regulators involved with the recent financial bubble and crash of 2008–9. Imitation and herding amplified events as the financial crisis evolved, providing to everyone involved the feeling of safety. Surely so many people could not be wrong!

A number of researchers are exploring the nature of human interactions in detail. Group-think and altruism emerge as important features. Psychologist Berns *et al.* (2010) have shown that anxiety arising from a difference in personal preferences motivates people to switch choices in the direction of consensus in a group context. Following the early studies by Nash (1951) with relevance to the prisoner's dilemma game, a number of people are exploring iterated games as models for human interactions. Using this approach, Vilone *et al.* (2012) have shown that social context is important and can determine whether consensus is reached. Iranzo *et al.* (2012) have recently shown how empathy and cooperation—where agents consider what might be acceptable before making offers and proposals to others—can emerge spontaneously and displace agents with other independent proposals. So there is no doubt that interactions between social agents exist and it must be investigated further, how far the methods of physics between atoms and molecules are applicable to social systems.

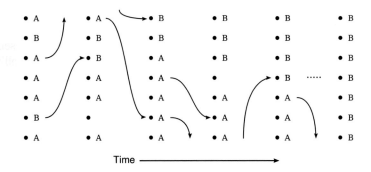

Figure 22.1 Schematic of the process of evolution of the distribution two different groups *A* and *B*, within a street. As a vacancy arises and one member vacates a house and moves out of the area, the house is reoccupied by a street resident in order to live adjacent to a fellow member of their group. The preferences *A*—*A* and *B*—*B* are greater than *A*—*B* or *B*—*A*. Such events can occur, even if the preferences are small and evolve slowly over long time periods (see text below) (Sketch after Schelling, 1969).

In 1969, the economist Thomas Schelling published his paper 'Models of Segregation' (Schelling, 1969) in which he discussed the issue of segregation and showed how it can result from discriminatory individual choices. In 1972 Weidlich published a physics paper on collective social behaviour, and Mimkes (1995, 2000, 2006b) proposed binary alloys as model of the multicultural society.

Schelling had no access to the tools of statistical physics, but he clearly understood how, by analogy, physical laws could be qualitatively implemented in the social sciences. He developed simple but profound models of social behaviour that showed phase transitions without ever using the term. The approach is simply illustrated; consider for example a community of two different groups *A* and *B* living in a street. If the populations of each community, N_A and N_B, are equal and there is no interaction or preference for association, then one could imagine the different members of the communities to be distributed randomly along the street. However, beginning with such a random distribution, Schelling showed, via a simple simulation method, that if a preference, no matter how small, exists for living alongside a family of similar religion, then movement, as house vacancies crop up, leads inevitably to a state of separation of the two groups (see Figure 22.1).

One of the current authors (PR) recalls over a 50-year period visiting an aunt who lived on a street in a small town in the north of England. In fact she had been born in the house during the 1920s. During the 1960s, as houses became vacant, they became occupied by Muslims from Pakistan who were moving into the area. Local people chose to move away. The aunt was, by the turn of the century, the only white person living in the street and when she died in 2007, a Muslim family bought the property. But at the funeral it was abundantly clear that she was not isolated in the community. Many of her Muslim neighbours came to what was a very traditional Christian funeral to pay their respects. There was no evidence of any antagonism towards non-Muslims from this local community.

It should be obvious that the ideas of Schelling are not just applicable to religious or social groups. They apply to any situation where agents are faced with binary or other more complicated choices. 'The preference to mix with people similar to oneself is as old as humanity' according to economist Paul Ormerod (2002). However, Schellings work demonstrates something even more interesting which is exemplified by the illustration of the previous paragraph. This is that pronounced or complete segregation follows even for the smallest of personal preferences.

A high degree of segregation in mixed communities is usually interpreted as being a bad thing by governments who assume it means strong antagonism between the groups. However, this type of modelling suggests this conclusion need not be so and might call into question the policies introduced by many western governments in recent years to promote integration. New thinking based on an examination of other parameters, such as the interaction 'potential' and dither that govern segregation should clearly be of greater value than simply monitoring the degree of segregation itself.

Missing from Schelling's approach is the concept of entropy and the idea of the dither. By including this, we now have two competing effects within the system. On the one hand we have interactions that lead to phase separation and, on the other hand, we have entropy that promotes diffusion of agents throughout the entire phase space of the system. The degree to which one or other of these effects dominate is controlled by the dither. If the noise induced by the dither is low, then the phase separation observed by Schelling will occur. However as the dither increase in importance, a point is reached when phase separation no longer occurs.

A concise description of the nature of social phases in binary systems, using a mean-field approximation developed for binary alloys—the so-called Bragg-Williams model named after the originators—has been described by Mimkes (2006b). For a binary system consisting of two species, A and B, the Lagrange function $\mathcal{L}$ is given by

$$\mathcal{L}(\lambda, N, x) = N\{(e_{AA} + x(e_{BB} - e_{AA}) + ex(1 - x)$$
$$-\lambda(x \ln x + (1 - x) \ln(1 - x))$$
$$+ \text{ terms independent of } x\}, \tag{22.1}$$

(cf. our discussion of optimisation problems in Section 19.3).

N is the total number of agents, $x = N_A/N$ is the fraction of type A, and $(1-x) = N_B/N$ is the fraction of type B. λ is the dither (or for binary alloys, the temperature).

As λ is a factor of entropy or disorder, it may be interpreted as a general 'tolerance' of disorder. In white societies, λ may be the general tolerance for black people within the society, in Muslim societies it may be the general tolerance for Christian people within the society, for shopkeepers, it may be the tolerance to accept foreign money.

The interaction between type A is e_{AA} and the interaction between type B is e_{BB}. The interactions between different types of agents A and B is e_{AB} and e_{BA}. Both parameters may be positive (sympathy), negative (antipathy) or zero (indifference). The sum of all interactions between different types of agents in this model rather surprisingly come together in the single parameter, $e = (e_{AB} + e_{BA}) - (e_{AA} + e_{BB})$.

In social systems, one might expect these parameters to change over time, albeit at very slow rates, over generations, rather than days, weeks or months, see for example Figure 22.2.

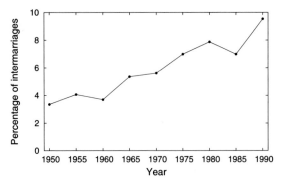

Figure 22.2 Intermarriage of Germans and non-Germans, expressed as a percentage of the total number of marriages in Germany, has steadily increased between 1950 and 1990 (Mimkes, 2000). This would correspond to a slow but steady change of interaction parameters in a Bragg-Williams type society.

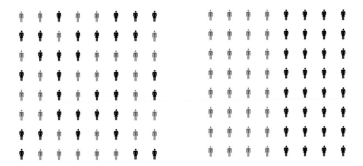

Figure 22.3 Results of Monte-Carlo computer simulation based on eqn (22.1). Left: Integrated binary society. Right: Segregated binary society for $e < 0$.

Change is driven by λ that controls the transitions between social states and societal dynamics. Whereas for a physical material $e_{AB} = e_{BA}$, this is not necessarily so for a social system. It follows that a Bragg-Williams society may exhibit a wider range of states spanning hierarchical, partnership, segregation, integration, aggression, and a so-called 'global' state, depending on the values of the various parameters.

For example, in both partnerships and hierarchies, the interaction $e_{AB} + e_{BA}$ between different agents is stronger than that between similar agents, $e_{AA} + e_{BB}$, resulting in $e > 0$. The Lagrangian may be thought of as a kind of 'happiness' function and is optimized when $x = 1/2$ and there are equal numbers of the two types of agents. If the numbers are not equal, clearly one will have some singletons either of type A or type B who cannot minimize their individual state by the virtue of being paired. For a partnership $e_{AB} = e_{BA}$, whereas in a hierarchy these parameters are not equal. This creates a difference within the pair itself.

Segregation occurs if $e < 0$ and both e_{AB} and e_{BA} are greater than zero, as is illustrated in Figure 22.3. Now the attraction between members of a group, $e_{AA} + e_{BB}$,

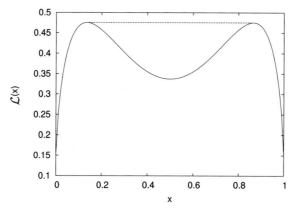

Figure 22.4 Lagrange function $\mathcal{L}(x)$ of a binary society with preference for the same groups, $e < 0$. The function shows two maxima for the minorities in each section. If x lies between the two maxima, the value of $\mathcal{L}(x)$ is given by the dashed line ('Maxwell construction').

is greater than that between members of different groups, $e_{AB} + e_{BA}$, as was the case in Schelling's model. Providing λ is less than a critical value, the Lagrangian typically takes the form illustrated in Figure 22.4, with two peaks.

The two maxima correspond to the optimal fractions of mixing of say Muslims in Christian areas and Christians in Muslim areas. If the actual fraction lies between either of these optimal values, the state is located on the common tangent as denoted by the dotted line or 'Maxwell construction' in Figure 22.4, giving the two-phase state a higher value than that predicted by the Lagrangian alone.

Maxwell and Gibbs first developed such constructions, for the understanding of phase separations in fluids and alloys, in the late nineteenth century. At equilibrium the derivative of the Lagrangian with respect to the fraction x is zero. If, for simplicity, we set $e_{AA} = e_{BB}$, we obtain from eqn (22.1) the result

$$\lambda = e(1 - 2x)/(\ln x - \ln(1 - x)). \tag{22.2}$$

For mixtures such as sugar in tea, or metal alloys, such a function yields the temperature consistent with a fraction x of dissolved material. If, at a particular temperature, one tries to dissolve more sugar into the tea, the sugar simply falls to the bottom of the cup. A higher temperature is needed to dissolve the excess. Figure 22.5 illustrates this behaviour with experimental data for gold platinum alloys.

Translating this into a social context, allows us to explore the solution or assim-ilation of different communities within societies. For example, Figure 22.6 shows the plot for intermarriage of Catholics and non-Catholics across Germany in 1991. Points on the parabolic curve correspond to points where both communities are integrated. However, above the 'integration limit' of 0.2, the data points depart from the curve. The roughly constant marriage rate of 33% corresponds to the equilibrium temper-ature, indicating a mutual religious tolerance of different neighbours. In states with a high percentage of Catholics, Mimkes found segregation into mainly Catholic and mainly non-Catholic areas. Making a comparison between Germany, Switzerland, and

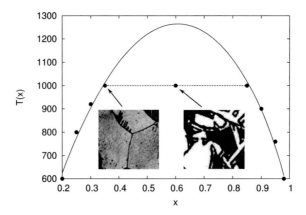

Figure 22.5 Phase diagram of gold platinum alloys (after Hansen, 1958). The curve $T(x)$ corresponds to the temperature that is needed to dissolve the percentage x of platinum. At $600°$ Celsius, gold may solve only 20% platinum. At $1000°$ Celsius the 'solubility limit' is $x = 0.35$ or 35% platinum. Any higher percentage will lead to segregation, as shown by the bright areas with mainly gold, and dark parts with mainly platinum (Data: Mimkes, 1995).

Northern Ireland in 1991, he found that Germany and Switzerland had similar characteristics, but for Northern Ireland he found that the data suggested a much lower integration temperature of about 2.3%. Such lower tolerance values arise from larger values for the repulsion, $e_{AB} + e_{BA}$, and these may give rise ultimately to aggression between the communities.

A similar picture has been obtained for intermarriage between African and non-African Americans in 33 states of the USA in 1988. As shown in Figure 22.7, intermarriage is ideal up to the 'solubility limit' of 1.1% African Americans in each state. For states with a higher percentage we find a constant portion of intermarriage of 1.1%, indicating the 'social equilibrium temperature' between different states of the USA. The 'solubility limit' leads again to segregation into predominate white and predominantly African American ghetto areas.

Mimkes further suggests that other aspects of societal change arise as the standard of living increases. This corresponds to thermodynamics, where molecules move only collectively in solids. With rising temperature, beyond the melting point, the solid will change into the liquid state, where molecules can move individually. In the same way, people in a hierarchy may only behave in a collective manner. With rising standard of living λ, the hierarchy will change into a democracy, where people can act individually. For example, at the end of the twentieth century, the regimes in Spain, Portugal, Greece, the USSR, and Eastern Europe, as well as in Brazil, Argentina, and Uruguay gave way to democracies due to the rise in standard of living. If we look at the standard of living λ of all countries in the world, we find most countries way above $5000 US per capita are democratic, and way below $5000 US per capita, most countries are in a hierarchic rule. Will Islamic states also evolve from the hierarchical rule of clerics to more democratic rule, as living standards increase? These arguments suggest they will; time will tell.

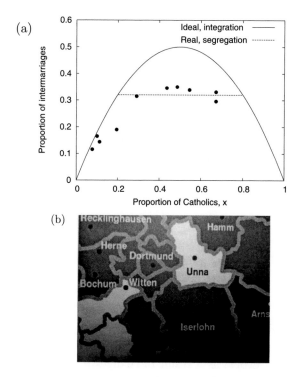

Figure 22.6 (a) Intermarriage between Catholics and non-Catholics in ten different states of Germany in 1991. Intermarriage is ideal up to the 'integration limit' of $x = 0.2$ or 20% Catholics in any state of Germany. For states with a higher percentage, we find segregation (Mimkes, 2001). (b) shows Westphalia with 40% Catholics segregating into black areas with mainly Catholics and white areas with mainly non-Catholics. Grey areas have equal distribution of black and white areas. (Statistisches Jahrbuch der Bundesrepublik Deutschland 1991–2001).

A number of other authors have taken over the model of magnetism and explored similar phenomena in society (Galam, 2008). Opinion formation and herding is one such effect that is amenable to analysis by these methods. More recently, historians and evolutionary biologists, such as Turchin (see Spinney, 2012) and physicists such as Ausloos (Ausloos and Petroni, 2007, Vitanov *et al.*, 2010) have begun to look at societal dynamics and the evolution of societies. In this way, history is examined as a science, rather than a series of stories and special events. Not everyone is comfortable with this and there is no doubt much more to be done in the way of refining the theory, but the approach does begin to offer new insights into the way we have evolved.

Karl Marx wrote in his *Thesen über Feuerbach* (1845): 'The philosophers have only interpreted the world, in various ways; the point is to change it.' It is usually true of politicians that they have a particular world-view and seek election by making promises to change the world in line with that viewpoint. However, as we can now see, economic and social systems are complex and we have barely scratched the surface in terms of

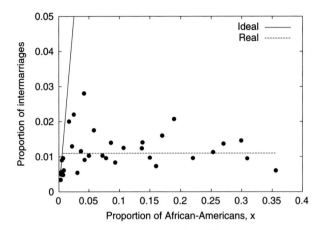

Figure 22.7 Intermarriage between African and non-African Americans in 33 states of the USA in 1988. Intermarriage is ideal up to the 'solubility limit' of $x = 0.011$ (1.1%) African Americans in each state. For states with a higher percentage we find a (roughly) constant portion of intermarriage of 1.1%, indicating the 'social equilibrium temperature', with some variation between different states of the USA. The 'solubility limit' leads again to segregation into white and predominant African American areas. (data: US Bureau of the Census, 1990.)

understanding how they evolve as a result of changes in rules and other parameters that control their evolution. So it is not surprising that politicians frequently fail in their quest.

Statistical physics and thermodynamics is one of the oldest of physical sciences, having been developed during the industrial revolution as a response to understanding gases and their use in steam engines. One could only speculate what might have been the situation in the social and economic sciences, if the theory had been available some years earlier when the French philosopher and founder of sociology, Auguste Comte (1798–1857) was in his prime. Today, the impact of the subject is clear from looking at the many new journals devoted to quantitative and mathematical sociology.

Yet the theory of Black and Scholes remains the theory of choice for many financial traders, when assessing options and derivatives. Empirical studies of detailed high frequency financial data not only demonstrate the point that this theory is inadequate and fails in ways that are extremely serious, they also points to new ways to understand the underlying fluctuations and price movements. Equally, mainstream economics seems still to be rooted in the paradigm of neoclassical equilibrium theory and the concept of the rational economic agent.

A few in the community (e.g. Bouchaud, 2008, Ormerod and Helbing, 2012) are pointing to the need for change in the way economics is taught in our universities. We hope that this book will help this aim and support those who recognize the subject is in need of a new paradigm; one that allows a more productive treatment of economic systems.

References

Abe, S., Beck, C., and Cohen, E.G.D. (2007). Superstatistics, thermodynamics, and fluctuations. *Physical Review E*, **76**, 031102 (5 pages).

Abramowitz, M. and Stegun, I.A. (1965). *Handbook of Mathematical Functions*. Dover Publications: New York.

Acheson, D.J. (1990). *Elementary Fluid Dynamics*. Oxford University Press.

Andersen, P.W. (1972). More is different. *Science*, **177**, 393–6.

Angle, J. (1986). The Surplus Theory of Social Stratification and the Size Distribution of Personal Wealth. *Social Forces*, **65**, 293–326.

Arnold, L. (2012). *Makroökonomik* (4th edn). Mohr Siebeck, Tübingen.

Aruka, Y. and Mimkes, J. (2011). An Evolutionary Theory of Economic Interaction: Introduction to Socio- and Econo-Physics. In: Aruka, Y. (ed.), *Complexities of Production and Interacting Human Behaviour*. Springer, Berlin Heidelberg. pp 113–27.

Ausloos, M. and Petroni, F. (2007). Statistical dynamics of religions and adherents. *Europhysics Letters*, **77**, 38002 (4 pages).

Bachelier, L. (1900). Théorie de la spéculation. *Annales Scientifiques de l'École Normale Supérieure*, **3**, 21–86.

Backhouse, R.E. (2002). *The Penguin History of Economics*. Penguin Books: London.

Ball, P. (2004). *Critical Mass—How one thing leads to another*. William Heineman: London.

Beck, C. and Cohen, E.G.D. (2003). Superstatistics. *Physica A*, **322**, 267–275.

Berg, J.E. and Rietz, T.A. (2006). The Iowa Electronic Markets: Stylized Facts and Open Issues. In: Hahn, R. and Tetlock, P.C. (eds.), *Information Markets: A New Way of Making Decisions in the Public and Private Sectors*. AEI Press: Washington D.C. pp. 142–69.

Berns, G., Capra, C.M., Moore, S., and Noussair, C. (2010). Neural Mechanisms of the Influence of Popularity on Adolescent Ratings of Music. *Neuroimage*, **49**, 2687–96.

Bernstein, P.L. (1996). *Against the Gods: The Remarkable Story of Risk*. John Wiley.

Black, F. and Scholes, M. (1973). The Pricing of Options and Corporate Liabilities. *Journal of Political Economy*, **81**, 637–54.

Bofinger, B. (2003). *Grundzüge der Volkswirtschaftslehre*. Pearsonstudium: München, Germany.

Borland, L. (2002a). Option Pricing Formulas Based on a Non-Gaussian Stock Price Model. *Physical Review Letters*, **89**, 098701 (4 pages).

Borland, L. (2002b). A theory of non-Gaussian option pricing. *Quantitative Finance*, **2**, 415–31.

Bouchaud, J.-P. and Cont, R. (1998). A Langevin approach to stock market fluctuations and crashes. *European Physical Journal B*, **6**, 543–50.

Bouchaud, J.-P. and Potters, M. (2000). *Theory of financial Risk and Derivative Pricing*. Cambridge University Press.

Bouchaud, J.-P. (2008). Economics needs a scientific resolution. *Nature*, **455**, 1181.

Burda, Z., Jurkiewicz, J., and Nowak, M.A. (eds.), (2005). *Conference on Applications of Random Matrices to Economy and Other Complex Systems, Cracow, Poland, May 25 - 28, 2005*, Jagellonian University Crakow, Poland. (Recognised by the European Physical Society Vol. 36 No. 9 Cracow APOBB B36(9)2603-2838 (2005)).

Caccioli, F., Marsili, M., and Vivo, P. (2009). Eroding market stability by proliferation of financial instruments. *European Physical Journal B*, **71**, 467–79.

Caldarelli, G. (2007). *Scale-Free Networks*. Oxford University Press.

Cartan, H. (2006). *Differential Forms*. Dover Books on Mathematics. Dover Publications.

Cercignani, C. (1998). *Ludwig Boltzmann: The man who trusted atoms*. Oxford University Press.

Champernowne, D.C. (1953). A Model of Income Distribution. *The Economic Journal*, **63**, 318–51.

Chatterjee, A., Yarlagadda, S., Chakrabarti, B.K. (eds.), (2005), *Econophysics of wealth distributions*. Springer.

Cleaver, T.C. (2011). *Economics - the basics* (2nd edn). Routledge, Taylor and Francis Group: London.

Coelho, R., Hutzler, S., Repetowicz, P., and Richmond, P. (2007a). Sector analysis for a FTSE portfolio of stocks. *Physica A*, **373**, 615–26.

Coelho, R., Gilmore, C.G., Lucy, B., Richmond, P., and Hutzler, S. (2007b). The evolution of interdependence in world equity markets—Evidence from minimum spanning trees. *Physica A*, **376**, 455–66.

Coelho, R., Richmond R., and Hutzler, S. (2008a). A random-matrix-theory-based analysis of stock markets from different countries. *Advances in Complex Systems*, **11**, 655–68.

Coelho, R., Richmond, P., Barry, J., and Hutzler, S. (2008b). Double power laws in income and wealth distributions. *Physica A*, **387**, 3847–51.

Coffey, W.T., Kalmykov, Y.P., and Waldron, J.T. (2004). *The Langevin Equation* (2nd edn). World Scientific.

Comte, A. (1856). *Social Physics: from the Positive Philosophy*. Calvin Blanchard: New York.

Dacorogna, M.M., Muller, U.A., Nagler, R.J., Olsen, R.B., and Pictet, O.V. (1993). A geographical model for the daily and weekly seasonal volatility in the FX market. *Journal of International Money and Finance*, **12**, 413–38.

Dahlke, R., Fakler, R.A., and Morash, R.P. (1989). A sketch of the history of probability theory. *Mathematics Education*, **5** 218–32.

Davis, M. and Etheridge, A. (2006). *Louis Bachelier's Theory of Speculation: The Origins of Modern Finance*. Princeton University Press.

Dillow, C. (2012). Risk: the simple and the complex. *Investors Chronicle*, 21–27 September 2012, page 19.

Drăgulescu, A.A. and Yakovenko, V.M. (2002). Probability distribution of returns in the Heston model with stochastic volatility. *Quantitative Finance*, **2**, 443–53.

Einstein, A. (1905). Über die von der molekularkinetischen Theorie der Wärme geforderte Bewegung von in ruhenden Flüssigkeiten suspendierten Teilchen. *Annalen der Physik*, **322**, 549–60.

Èrdi, P. (2008). *Complexity Explained*. Springer Verlag, Berlin, Heidelberg.

Flanders, H. (1990). *Differential Forms with Applications to the Physical Sciences*. Dover Publication, New York.

Feigenbaum, J.A. and Freund P.G.O. (1996). Discrete Scaling in Stock Markets Before Crashes. *International Journal of Modern Physics*, **12**, 57-60.

Fründ, T. (2002). *Analyse von Einkommen, Vermgen und Gesellschaft mit physikalischen Mitteln*. Staatsexamensarbeit, Fachbereich Physik, Universität Paderborn, Germany.

Gabaix, X., Gopikrishnan, P., Plerou, V., and Stanley, H.E. (2003). Understanding the cubic and half-cubic laws of financial fluctuations. *Physica A*, **324**, 1–5.

Galam, S. (2008). Sociophysics: a review of Galam models. *International Journal of Modern Physics C*, **19**, 409–40.

Gardiner, C.W. (2004). *Handbook of Stochastic Methods*. Springer Verlag, Berlin, Heidelberg.

Gell-Mann, M. and Tsallis, C. (eds.) (2004). *Nonextensive Entropy Interdisciplinary Applications*, Oxford University Press.

Georgescu-Roegen, N. (1971). *The entropy law and the economic process*. Harvard University Press: Cambridge MA, USA.

Gibrat, R. (1931). *Les Inégalites Économiques*. Libraire du Recueil Sirey: Paris.

Gopikrishnan, P., Meyer, M., Amaral, L.A.N., and Stanley, H.E. (1998). Inverse Cubic Law for the Probability Distribution of Stock Price Variations. *European Physical Journal B: Rapid Communications*, **3**, 139–40.

Gopikrishnan, P., Plerou, V., Amaral, L.A.N., Meyer, M., and Stanley, H.E. (1999). Scaling of the Distributions of Fluctuations of Financial Market Indices. *Physical Review E*, **60**, 5305–16.

Halley, E. (1693). An estimate of the degrees of mortality of mankind, drawn from curious tables of the births and funerals at the city of Breslaw, with an attempt to ascertain the price of annuities on lives. *Philosophical Transactions*, **17**, 596–610.

Hansen, M. (1958). *Constitution of binary alloys*. McGraw–Hill, New York.

Hardiman, S.J., Richmond, P., and Hutzler, S. (2009). Calculating statistics of complex Networks through random walks with an application to the on-line social network Bebo. *European Physical Journal B*, **71**, 611–22.

Hardiman, S.J., Richmond P., and Hutzler, S. (2010). Long-range correlations in an online betting exchange for a football tournament. *New Journal of Physics*, **12**, 105001 (14 pages).

Hardiman, S.J., Tobin, S.T., Richmond, P., and Hutzler, S. (2011). Distributions of certain market observables in an on-line betting exchange. *Dynamics of Socio-Economic Systems*, **2**, 121–37.

Heston, S.L. (1993). A Closed-Form Solution for Options with Stochastic Volatility with Applications to Bond and Currency Options. *The Review of Financial Studies*, **6**, 327–43.

Hull, J.C. (2006). *Options, Futures and other Derivatives* (6th edn.). Pearson Prentice Hall.

Hutzler, S., Delaney, G., Weaire, D., and MacLeod, F. (2004). Rocking Newton's cradle. *American Journal of Physics*, **72**, 1508–16.

Iranzo, J., Floría, L.M., Moreno, Y., and Sánchez, A. (2012). Empathy Emerges Spontaneously in the Ultimatum Game: Small Groups and Networks. *PLOS ONE*, **7**, e43781 (8 pages).

Jaynes, E.T. (1991). How should we use entropy in economics? (unpublished manuscript).

Johansen, L. (1959). Substitution versus Fixed Production Coefficients in the Theory of Economic Growth: A Synthesis. *Econometrica*, **27**, 157–76.

Johansen, A., Sornette, D., and Ledoit, O. (1999). Predicting Financial Crashes Using Discrete Scale Invariance. *Journal of Risk*, **1**, 5–32.

Johansen, A., Sornette, D., and Ledoit, O. (2000). Crashes as critical points. *International Journal of Theoretical and Applied Finance*, **3**, 219–55.

Khintchine, A.Y. and Lévy, P. (1936). Sur les lois stable. *Comptes rendus de l'Académie des sciences Paris*, **202**, 374–6.

Kümmel, R. (2011). *The second Law of Economics, Energy, Entropy and the Origin of Wealth*. Springer, Heidelberg.

Lax, M., Cai, W., and Xu, M. (2006). *Random Processes in Physics and Finance*. Oxford University Press.

Liu, Y., Gopikrishnan, P., Cizeau, P., Meyer, M., Peng, C.-K., and Stanley, H.E. (1999). The statistical properties of the volatility of price fluctuations. *Physical Review E*, **60**, 1390–400.

Lobato, I.N. and Velasco, C. (2000). Long Memory in Stock-Market Trading Volume. *Journal of Business and Economic Statistics*, **18**, 410–27.

Malcai, O., Biham, O., Richmond, P., and Solomon, S. (2002). Theoretical analysis and simulations of the generalized Lotka-Volterra model. *Physical Review E*, **66**, 031102 (6 pages).

Mandelbrot, B. (1960). The Pareto-Lévy law and the distribution of income. *International Economic Review*, **1**, 79–106.

Mandelbrot, B. (1963). The variation of certain speculative prices. *Journal of Business*, **36**, 394–419.

Mandl, F. (1998). *Statistical Physics* (2nd edn.). John Wiley & Sons.

Mantegna, R.N. and Stanley, H.E. (1995). Scaling behaviour in the dynamics of an economic index. *Nature*, **376**, 46-9.

Mantegna, R.N. and Stanley, H.E. (2000). *An Introduction to Econophysics*. Cambridge University Press.

Markowitz, H.M. (1959). *Portfolio Selection: Efficient Diversification of Investments.* John Wiley & Sons, New York.

McCauley, J.L. (2004). *Dynamics of Markets: Econophysics and Finance.* Cambridge University Press.

Merton, R.C. (1973). Theory of Rational Option Pricing. *Bell Journal of Economics and Management Science* (The RAND Corporation), **4**, 141–83.

Mimkes, J. (1995). Binary Alloys as a Model for Multicultural Society. *Journal of Thermal Analysis*, **43**, 521–37.

Mimkes, J. (2001). Die familiale Integration von Zuwanderern und Konfessionsgruppen – Zur Bedeutung von Toleranz und Heiratsmarkt. In: Thomas Klein (ed.), *Partnerwahl und Heiratsmuster*. Leske und Budrich, Opladen, 233–62.

Mimkes, J. (2000). Society as a many-particle system. *Journal of Thermal Analysis*, **60**, 1055–69.

Mimkes, J.A. (2006a). A thermodynamic formulation of economics. In: Chakrabarti, B., Chakraborti, A., and Chatterjee A. (eds.), *Econophysics and Sociophysics: Trends and Perspectives*. Wiley VCH, 1–34.

Mimkes, J.A. (2006b). A thermodynamic formulation of social sciences. In: Chakrabarti, B., Chakraborti, A., and Chatterjee A. (eds.), *Econophysics and Sociophysics: Trends and Perspectives*. Wiley VCH, 279–310.

Mimkes, J. (2006c). Concepts of Thermodynamics in Economic Growth. In: Namatame A., Kaizouji, T., and Aruka, Y. (eds.), *The Complex Networks in Economic Interactions. Lecture Notes in Economics and Mathematical Systems*. Springer: Berlin, Heidelberg.

Mimkes, J. (2010a). Putty and Clay – Calculus and Neoclassical Theory. *Dynamics of Socio-Economic Systems*, **2**, 1-8.

Mimkes, J. (2010b). Stokes integral of economic growth: Calculus and the Solow model. *Physica A*, **389**, 1665–76.

Mimkes, J. (2012). Introduction to Macro-Econophysics and Finance. *Continuum Mechanics and Thermodynamics*, **24**, 731–7.

Morris, H.M. (1982). *Men of Science, Men of God: Great Scientists Who Believed the Bible*. Master Books, El Cajon, CA, USA.

Müller, L. (2001). *Handbuch der Elektrizittswirtschaft*. Springer Verlag.

Nash, J. (1951). Non-Cooperative Games. *The Annals of Mathematics*, **54**, 286–95.

Nicolis, G. and Nicolis, C. (2007). *Foundations of complex systems*. World Scientific.

Ormerod, P. (2002). Sense on segregation. *Prospect magazine*, January 2002.

Ormerod, P. and Helbing, D. (2012). Back to the Drawing Board for Macroeconomics. In: Coyle, D. (ed.), *What's the Use of Economics?: Teaching the Dismal Science After the Crisis*. London Publishing Partnership, pp 131–51.

Pareto, V. (1897). *Cours d'Économie Politique*. Librairie Droz (Genève), new edition of the original, 1964.

Phelps, E.S. (1963). Substitution, Fixed Proportions, Growth and Distribution. *International Economic Review*, **4**, 265–88.

Plerou, V., Gopikrishnan, P., Rosenow, B., Nunes, L.A., Guhr, T., and Stanley H.E. (2002). Random matrix approach to cross correlations in financial data. *Physical Review E*, **65**, 066126 (18 pages).

Preis, T. and Stanley, H.E. (2010). Switching Phenomena in a System with No Switches. *Journal of Statistical Physics*, **138**, 431–46.

Queirós, S.M.D., Moyano, L.G., de Souza, J., and Tsallis, C. (2007). A nonextensive approach to the dynamics of financial observables. *European Physical Journal B*, **55**, 161–7.

Repetowicz, P., Hutzler, S., and Richmond, P. (2005). Dynamics of money and income distributions. *Physica A*, **356**, 641–54.

Richmond, P. and Solomon, S., (2001). Power Laws are Disguised Boltzmann Laws. *International Journal of Modern Physics C*, **12**, 333–43.

Richmond, P., Hutzler, S., Coelho, R., and Repetowicz, P. (2006). A Review of Empirical Studies and Models of Income Distributions in Society. In: Chakrabarti, B.K., Chakraborti, A. and Chatterjee, A. (eds.) *Econophysics and Sociophysics: Trends and Perspectives*. Wiley-VCH, Berlin, pp 129–58.

Richmond, P. (2007). A Roof over your Head; House Price Peaks in the UK and Ireland. *Physica A*, **375**, 281–7.

Richmond, P. (2009). Will house prices rise in 2007? A comparative assessment of house prices in London and Dublin. In: Faggini, M. and Lux, T. (eds.), *Coping with the Complexity of Economics*. Springer.

Richmond, P. and Roehner, B. (2012). The predictable outcome of speculative house price peaks. *Evolutionary and Institutional Economics Review*, **9**, 125–40.

Roehner, B.M. (1999). Spatial analysis of real estate price bubbles: Paris, 1984-1993. *Regional Science and Urban Economics*, **29**, 73–88.

Roehner, B.M. (2000). Identifying the bottom line after a stock market crash. *International Journal of Modern Physics*, **11**, 91-100.

Roehner, B.M. (2001). *Hidden Collective Factors in Speculative Trading*. Springer Verlag, Berlin.

Roehner, B.M. (2006). Real Estate Price Peaks – A Comparative Overview. *Evolutionary and Institutional Economics Review*, **2**, 167–82.

Roehner, B.M. (2007). *Driving forces in physical, biological and socio-economic phenomena*. Cambridge University Press.

Roman, H.E., Porto, M., and Dose, C. (2008). Skewness, long-time memory, and non-stationarity: Application to leverage effect in financial time series. *Europhysics Letters*, **84**, 28001 (5 pages).

Romer, D. (1996). *Advanced Macroeconomics*. McGraw-Hill.

Sabatelli, L., Keating, S., Dudley, J., and Richmond, P. (2002). Waiting Time Distributions in Financial Markets. *European Journal of Physics B*, **27**, 273–5.

Schelling, T.C. (1969). Models of Segregation. *American Economic Review*, **59**, 488–93.

Schumpeter, J.A. (1954). *History of Economic Analysis*. Oxford University Press: New York.

Shen, J. and Zheng, B. (2009). On return-volatility correlation in financial dynamics. *Europhysics Letters*, **88**, 28003 (6 pages).

Shirras, G.F. and Craig, J.H. (1945). Sir Isaac Newton and the Currency. *The Economic Journal*, **55**, 217–41.

Shaw, W.A. (1896). *Select Tracts and Documents of English Monetary History 1626-1730*. London Clement Wilson. (Reprinted George Harding, 1935).

Sinha, S., Chatterjee, A., Chakraborti, A., and Chakrabarti, B.K. (2011). *Econophysicsi – An introduction*. Wiley-VCH.

Silva, A.C. and Yakovenko, V.M. (2003). Comparison between the probability distribution of returns in the Heston model and empirical data for stock indexes. *Physica A*, **324**, 303–10.

Slanina, F. (2004). Inelastically scattering particles and wealth distribution in an open economy. *Physical Review E*, **69**, 046102 (7 pages).

Solomon, S. and Richmond, P. (2001). Stability of Pareto-Zipf law in non-stationary economies. In: Kirman, A. and Zimmermann, J.B. (eds.), *Economies with heterogeneous interacting agents*. Springer-Verlag, pp 141–59.

Solomon, S. and Richmond, P. (2002). Stable Power Laws in Variable Economies; Lotka-Volterra implies Pareto-Zipf. *European Physical Journal B*, **27**, 257–261.

Sornette, D. and Johansen, A. (1997). Large financial crashes. *Physica A*, **245**, 411–22.

Sornette, D. (2003). *Why stock markets crash; critical events in complex financial systems*. Princeton University Press.

Sornette, D. and Woodard, R. (2010). Financial Real Estate and Derivative bubbles. In: Takayasui M., Watanabe, T., and Takayasu, H. (eds.), *Econophysics approaches to large scale business data and financial crisis*. Springer, pp 101–48.

Takayasu, M., Watanabe, T., and Takayasu, H. (eds.) (2010). *Econophysics approaches to large scale business data and financial crisis*. Springer Verlag.

Taylor, J. (1955). Copernicus on the Evils of Inflation and the Establishment of a Sound Currency. *Journal of the History of Ideas*, **16**, 540–7.

Touchette, H. and Beck, C. (2005). Asymptotics of superstatistics. *Physical Review E*, **71**, 016131 (6 pages).

Tsallis, C. (1988). Possible generalization of Boltzmann-Gibbs statistics. *Journal of Statistical Physics*, **52**, 479–87.

Spinney, L. (2012). Human cycles; History as Science. *Nature*, **488**, 24–6.

Uhlenbeck, G.E. and Ford, G.W. (1963). Lectures in Statistical Mechanics. In: *Lectures in Applied Mathematics, vol. I*, American Mathematical Society, Library of Congress catalogue card number 62-21480.

Vilone, D., Ramasco, J.J., Sánchez, A., and San Miguel, M. (2012). Social and strategic imitation: the way to consensus. *Scientific Reports*, **2**, Article number: 686 (7 pages).

Vitanov, N.K., Dimitrova, Z.I., and Ausloos, M. (2010). Verhulst-Lotka-Volterra (VLV) model of ideological struggle. *Physica A*, **389**, 4970–80.

Voit, J. (2001). *The Statistical Mechanics of Financial Markets*. Springer-Verlag, Berlin; Heidelberg, New York.

Weaire, D. and Hutzler, S. (1999). *The Physics of Foams*. Clarendon Press, Oxford.

Weaire, D. and Hutzler, S. (2009). Foam as a complex system. *Journal of Physics: Condensed Matter*, **21**, 474227 (4 pages).

Weidlich, W. (1972). The use of stochastic models in sociology. *Collective Phenomena*, **1**, 51–9.

Wilson, K.G. (1979). Problems in physics with many length scales. *Scientific American*, **241**, 158–179.

Woolfson, M.M. and Woolfson, M.S. (2007). *Mathematics for Physics*. Oxford University Press.

Yakovenko, V.M. and Silva, A.C. (2005). Two-class structure of income distribution in the USA: Exponential bulk and power-law tail. In: Chatterjee, A., Yarlagadda, S., and Chakrabarti, B.K. (eds.), *Econophysics of Wealth Distributions*, Springer series 'New Economic Windows', 15–23.

Zhou, W.-X. and Sornette, D. (2006). Is There a Real Estate Bubble in the US? *Physica A*, **361**, 297–308.

Index